Woodworking 101

The Taunton Press
Inspiration for hands-on living®

The Taunton Press, Inc., 63 South Main Street, PO Box 5506, Newtown, CT 06470-5506
e-mail: tp@taunton.com

Editor: Alex Giannini
Copy editor: Candace B. Levy
Indexer: Barbara Mortenson
Cover design: Teresa Fernandes
Interior design: Barbara Balch and Cathy Cassidy
Layout: Cathy Cassidy
Illustrators: Chuck Lockhart pp. 26–119, 232–295; Melanie Powell pp. 120–231
Photographers: Front cover (clockwise from left): Del Brown; Sloan Howard; Sloan Howard; Back cover Del Brown;
Philip Dutton pp. 2–93; Matthew Teague pp. 94–119; Slaon Howard pp. 120–231; Del Brown pp. 232–295

Fine Woodworking® is a trademark of The Taunton Press, Inc., registered in the U.S. Patent and Trademark Office.

The following names/manufacturers appearing in *Woodworking 101* are trademarks:
Forstner®, Kevlar®, Masonite®, Nicholson®, Phillips®, Quick-Grip®, Spirograph®, Super Bowl®,
Super Glue®, Tapcon®, Watco®

Library of Congress Cataloging-in-Publication Data
Fraser, Aimé Ontario.
  Woodworking 101 : skill-building projects that teach the basics / Aimé Fraser, Matthew Teague and Joe Hurst-Wajszczuk.
    pages cm
  ISBN 978-1-60085-368-5 (pbk.)
1. Furniture making--Amateurs' manuals. I. Teague, Matthew. II. Hurst-Wajszczuk, Joe. III. Title. IV. Title: Woodworking
one-o-one.
  TT195.F73 2012
  684'.08--dc23
                    2011049154

Printed in the United States of America
10 9 8 7 6 5 4 3 2 1

About Your Safety: Working wood is inherently dangerous. Using hand or power tools improperly or ignoring safety
practices can lead to permanent injury or even death. Don't try to perform operations you learn about here (or elsewhere)
unless you're certain they are safe for you. If something about an operation doesn't feel right, don't do it. Look for another
way. We want you to enjoy the craft, so please keep safety foremost in your mind whenever you're in the shop.

Woodworking 101

Includes Step-by-Step Instructions for 7 Projects

Aimé Fraser, Matthew Teague, and Joe Hurst-Wajszczuk

The Taunton Press

Contents

SETTING UP YOUR SHOP

WOODWORKING PROJECTS

Your First Shop

Though they share many operations, woodworking and carpentry are different trades. Carpenters work on site with a tool kit chosen for versatility and mobility. Wood-workers work at a bench in a shop (though they may share it with a car or hot-water heater) and use a larger collection of specialized tools capable of a wider range of operations and closer tolerances.

This section starts you out with the essentials—the tools you'll need to accomplish the most basic operations.

What to Consider

As you progress in woodworking, you'll find that there are some important differences between woodworking and carpentry tools. But while you're still learning basic skills, you can use the same tools a carpenter uses. They're readily available and less expensive, so you can get started in woodworking without a big outlay for tools.

Laying a Foundation

Your first tool should be the one that most clearly declares you a woodworker—a sturdy workbench. Determining the right bench is a delicate business, and many advanced woodworkers spend huge amounts of time obsessing over and building their perfect bench. New woodworkers need some experience before they can evaluate their bench needs, so the goal is to get something that meets the basic criteria without much fuss. Later, you can modify or retire it.

You'll also need a means of getting workpieces out of the lumber you buy, and a circular saw fits the bill. It's not elegant, but with the right setup, it does a good job with both solid wood and sheet goods. You'll need measuring tools for cutting out those parts and

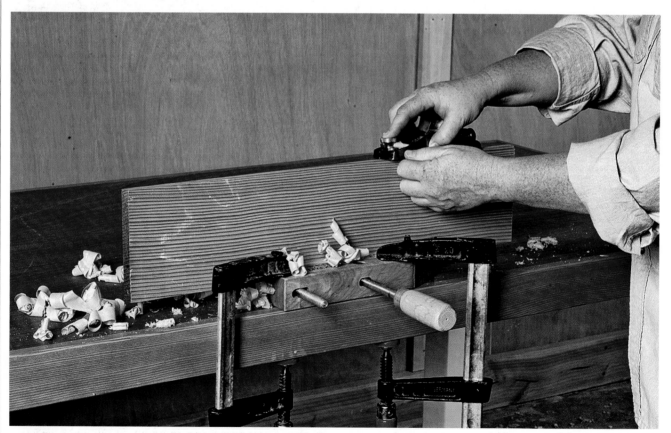

With a good collection of clamps, you can hold the work to the workbench without the need for a vise. Eventually, you'll want a vise for the convenience of quickly securing and releasing parts.

In the beginning, before you've bought many tools, **your shop space can be simple. You may even have room to share it with the car.**

for getting them square, planes and chisels for fitting the pieces, and clamps to hold it all together while you fasten the joints using hammers, screwdrivers, and a cordless drill/driver. Once your projects are complete, sanding tools are used to smooth them before finishing. To keep you safe and efficient all the while, be sure to have a few cleaning tools and some proper protection devices.

One Step at a Time

As you take on new projects and get familiar with your tools, you'll be learning new skills—how to stand when using the router, how to drive a screw with a cordless drill/driver, and how to plane. Then you'll learn related skills—how to set up guides for the saw, how to sharpen

your planes, and how to measure and mark efficiently and without error. As you do, you'll be learning how to think like a woodworker. You'll be learning the correct order in which to tackle operations, how to reduce mistakes, how to work efficiently, and more.

You're embarking on a craft that takes a lifetime to master. There's a lot going on, and it's important that you give yourself room for imperfection. Allot more time and materials than you think you'll need, and when something goes wrong, make the most of it. Before rushing off for the do-over, figure out just what went wrong and what you might have done to prevent it. You won't make that mistake again. With each one, you build up your store of woodworking wisdom. With each one, you get better.

Workbench

The bench is the heart of any woodworking shop. From laying out and gluing up to planing, shaping, and finishing, it's where work gets done. The problem for the beginning woodworker is figuring out just what a good bench is. If you look to current woodworking literature for help, you'll see all kinds of benches. One author says you must have feature X, another suggests only Y. It can be overwhelming, especially when you don't have the experience to evaluate a design. That's why I suggest you set up your first shop with a simple bench you can modify to suit your work style as your skills increase.

What to Buy

Every good woodworking bench, no matter what it looks like, meets a few basic criteria. First, it's sturdy enough to handle vigorous operations without moving. The legs are beefy, solidly joined, and well braced to anchor a thick, heavy top. If the bench lives in a shop

with uneven floors, it needs large-diameter leveling glides. A convenient working height is between 33" and 36" tall, and the top should be between 20" and 32" wide—wide enough to hold workpieces, but not so wide you can't reach across it. Most benches range from 4' to 7' long, but length is not a critical issue.

Though it's possible to build light, stiff tabletops, you want weight in a bench. The traditional top is at least 2¼" thick, laminated from tough beech or maple. A similar thickness layered from manufactured materials like

WHAT A BENCH CAN DO

■ BASE FOR CLAMPING

Assemble panels on the benchtop.

■ ASSEMBLY STATION

Use it as a steady base for applying edging.

■ FLAT, LEVEL WORK SURFACE

A bench provides a clear, stable surface for sanding.

plywood, medium-density fiberboard (MDF), and core doors works nearly as well and is much easier to build. Just be sure that the top is flat so you're able to quickly clamp anywhere around the periphery and underside of the benchtop. That means no lengthwise stretchers or drawers near the top.

Build or Buy?

Because your bench will be one of your first acquisitions when setting up shop, be conservative and choose a simple, sturdy bench that an inexperienced woodworker with minimal tools can build.

Fashion a beefy base from well-chosen dry lumber or simply buy sturdy metal legs. Use a store-bought laminated slab for the top or build your own from plywood and MDF. The cost of building is moderate in time and money, and you can hardly go wrong. Even if you build another bench down the road, you can always use a sturdy horizontal.

If you want to buy a simple bench, check with an industrial supply house. It won't look like a woodworking bench, but it'll do the job and cost less than building one. And though I can offer a strong argument for working your way up to a traditional cabinetmaker's bench,

A portable vise can be easily clamped in place to hold small complex workpieces at a convenient height for fine work.

there's nothing to stop you from buying one from the start if you choose. Read ahead to the part on benches in the next chapter to learn more about your choices before buying.

■ **ANCHOR JIGS**

Use a hook to hold boards for crosscutting.

■ **STABILIZE WORK**

A corner can serve as a vise for sanding or light planing.

Cordless Drill/Driver

Cordless drill/drivers are light enough to use anywhere, have no power cord to tangle, and have the ability to drive screws without breaking them—no wonder corded drill sales are down. A wonderfully versatile tool, your cordless drill/driver will likely end up being the most-used power tool in your shop.

What to Buy

Take the time to find a drill/driver that feels good in your hands. Balance, switch locations, and grip size are matters of taste, but they are more important to your satisfaction with the tool than motor ratings or foot-pounds of torque.

When selecting your drill/driver, don't be tempted to get the biggest one you can find. Stick with a 12- or 9-volt machine. They are lighter and easier to handle, and they have plenty of power to manage the jobs you do as a woodworker.

Two features that differentiate the cordless drill/driver from the mere drill are the high/low speed switch and the adjustable clutch. As a rule, use the low speed for driving screws and boring large holes. High speed is for drilling small holes and removing screws efficiently.

The adjustable clutch disengages the drive when the torque reaches a certain level, so even though the trigger is fully depressed, the bit won't turn. You can set this feature to drive screws flush with the surface, deep below the surface, or anywhere between.

WHAT A DRILL/DRIVER CAN DO

■ BORE HOLES

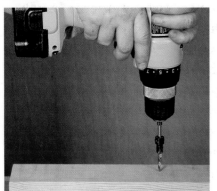

Use light pressure and stay in line with the hole.

■ DRILL WITH PRECISION

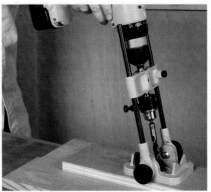

Jigs allow you to drill a hole at an exact angle—or location.

■ SAND WITH ATTACHMENTS

Disks and sleeves turn a drill into a sander.

Cordless Drill/Driver and Drill Bits

You'll use your cordless drill/driver more than any other power tool in the shop. Make sure your first has the features to satisfy your woodworking needs for years to come.

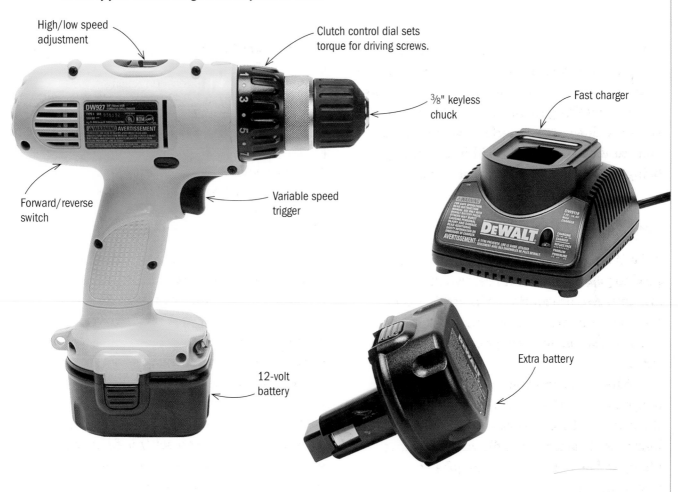

High/low speed adjustment

Clutch control dial sets torque for driving screws.

3/8" keyless chuck

Fast charger

Forward/reverse switch

Variable speed trigger

12-volt battery

Extra battery

■ CONSTRUCT STRONG JOINTS

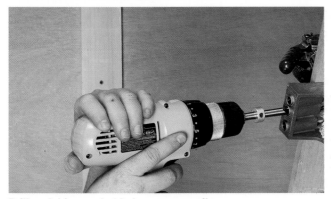

Drill and drive pocket-hole screws easily.

You'll also want to get a drill/driver that has a reverse switch located conveniently near the trigger. You'll need two batteries to handle big jobs and a charger that can do its job in an hour or less so you're not stranded in the middle of a job. For best battery life, charge your batteries when you note reduced performance but before they're fully dead.

Set the speed to high, set the clutch in drill position, and fully depress the trigger. That's the gist of boring holes, but boring good, clean holes requires that you attend to a few other factors.

Drill Bits

Whether you're drilling holes, driving screws, or using a specialized jig, there's a bit or drill to handle the job.

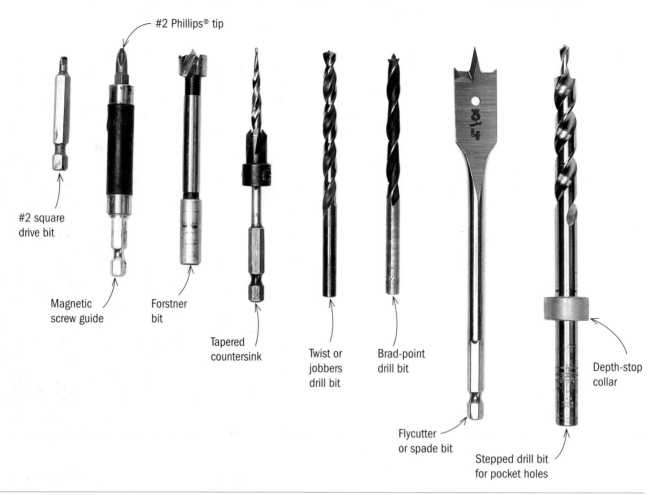

#2 Phillips® tip

#2 square drive bit

Magnetic screw guide

Forstner bit

Tapered countersink

Twist or jobbers drill bit

Brad-point drill bit

Flycutter or spade bit

Stepped drill bit for pocket holes

Depth-stop collar

Don't lean on the drill while boring a hole—a sharp bit requires only a firm, steady push. With a little experience, you'll be able to hear a change in pitch when the drill is close to cutting through. At that point, lighten up and let the drill do the work.

If you push too hard through the last bit of wood, you'll tear an ugly splinter from the back. For further insurance against tearout, hold or clamp the workpiece to a scrapwood backer board.

Choosing the right drill bit is crucial. A standard twist drill has a tendency to wander as it starts to cut, and the hole often winds up in the wrong place. You can reduce this tendency by punching a small dent or hole to direct the tip, but a better solution is to use brad-point bits designed for drilling wood. The sharp, protruding tip keeps the bit on course, and the bit's cutting angle helps sever wood fibers for clean, accurate cuts.

Carpenters, electricians, and other tradespeople commonly use a spade bit or flycutter in studs and joists. These inexpensive, disposable bits have their place, but Forstner® bits are a better choice for fine work. They leave a clean, smooth surface because the outer rim scores the surface while the inner part is sliced clean.

A hidden fastener. A pocket-hole screw joint is strong because the angled screw bites into long grain instead of weak end grain. The stepped drill used with pocket-hole jigs cuts a pilot hole for the screw shank and a shoulder to seat and conceal the screw head.

They're your best choice for smooth, large, or angled holes and for holes drilled to partial depth in a board.

To get started, buy a set of moderately priced twist drill bits in 1/64" increments up to 1/2". Beware of cheap sets, because the drill bits are often a little smaller than the measurement marked on their shanks. You'll also want brad-point bits—look for a set of at least seven. You can get by with an inexpensive set, but the more costly versions have a better tip design and cut more cleanly. For drilling pilot holes for screws, get at least five tapered countersink drill bits for the most common screw sizes.

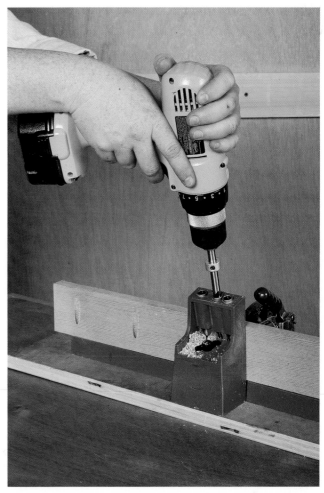

One joint, many uses. Using a pocket-hole jig, you can join at right angles, edge join, and even fasten good-looking bevels.

Finally, you'll want a set of Forstner bits in the common small sizes.

Driving Screws

The key to successful screw driving is to remember that you must keep the bit aligned with the centerline of the screws and fully engaged. If you aren't positioned behind the drill and pushing hard, torque forces the bit upward and tears up the screw head.

Ideally, the screw pushes the wood out of its way as it goes in, but in dense hardwoods or near the ends of a board the wood can split. Prevent this by drilling a pilot hole slightly narrower than the screw's thinnest part.

Edge Tools

Planing is one of my favorite things about working with wood. There's nothing as pleasant as the gentle, rhythmic exercise of guiding a plane over a board, the tearing-silk sound of a sharp blade on wood, the shimmering grain revealed, and the fragrant shavings spiraling to the floor. It's a defining act of woodworking, and knowing how to do it well is a fundamental skill.

A well-tuned plane can remove shavings as thin as four ten-thousandths of an inch. With such control, it's easy to get joints that fit perfectly, something you'd be hard pressed to do by setting up a powerful machine by trial and error. A plane can flatten a panel, profile an edge, and smooth a rough board until it feels like polished glass.

You'll also need to know how to use a chisel, no matter how many machines you end up with. Use it for chopping out waste wood in lap joints or dovetails, or to pare tissue-paper-thin shavings to fit a joint. When you know how to wield a chisel, you'll have the ability to join or shape wood any way you want.

WORK SMART

- If you drop your plane, check the edges for dings and file them out so they won't scratch the surfaces you plane.
- A plane laid on its side is exposed to bumps, dings, and stray hands—better to gently set it blade down on a slip of wood.
- It's deceptively easy to cut yourself on a sharp tool—always know where the cutting edge is in relation to your hands.

Learning to use planes and chisels is a lot like learning to putt or play the piano. It's a physical skill that takes practice to master. No new golfer expects to be an expert putter without putting in some practice time. It's the same with mastering planes and chisels. Spend some time practicing.

Before you can get anywhere with your edge-tool practice, you'll need to know how to sharpen. You'll never learn proper tool use or form with a dull tool—it fights you at every

Continued on p. 14

WHAT EDGE TOOLS CAN DO

■ FLATTEN AND SMOOTH

Remove saw marks and level edges after gluing.

■ PROFILE EDGES

Rounding or chamfering can be done by hand.

Block Plane and Butt Chisels

A small, one-handed block plane should be your first edge tool purchase; then buy a set of carpenter's butt chisels. Add a jig to hold your blades steady when honing, and you're ready to get to work (see "Sharpening" on p. 15).

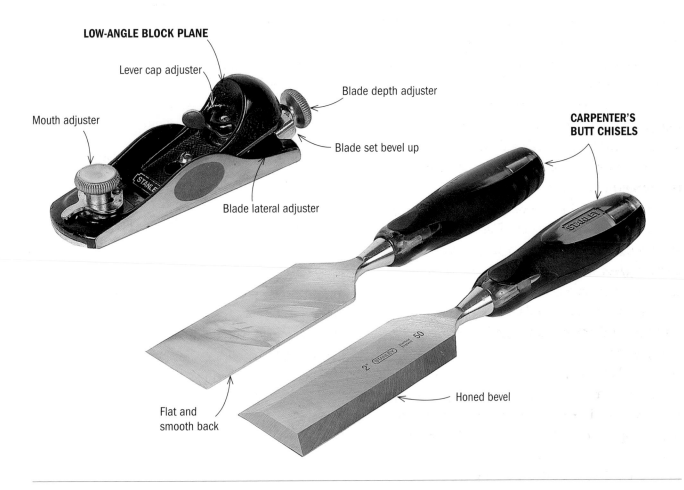

LOW-ANGLE BLOCK PLANE

Lever cap adjuster

Blade depth adjuster

Mouth adjuster

CARPENTER'S BUTT CHISELS

Blade set bevel up

Blade lateral adjuster

Honed bevel

Flat and smooth back

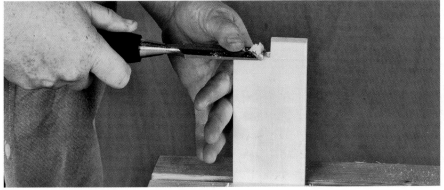

■ **PARE FINE SHAVINGS**

Remove tissue-thin slices for well-fitted joints.

■ **CHOP**

Remove waste with chisel and mallet.

Sharp tools make you a better woodworker. **Sharpen after about 15 minutes of continuous use or when it's no longer fun.**

What to Buy

Your first plane should be a low-angle block plane with an adjustable mouth, readily available at hardware stores and home centers. These are the easiest to set up and maintain, and they are more versatile than their simpler cousins. You can open the mouth, or gap between the blade and the sole of the plane, to accommodate thick chips for rough work or close it down for fine shavings. Be sure to get a plane with a lateral adjuster so you can keep the blade parallel to the sole.

Start with a few carpenter's butt chisels. They're shorter and beefier than cabinet-maker's chisels and designed for hitting with a hammer. It makes sense to buy your first chisels as a set. Make sure that whatever you buy has at least ⅜", ½", and ¾" chisels. Round out your set with the widest chisel you can get—1½" or even 2".

You'll need a honing jig for holding the blade steady at the correct angle when sharpening. Which jig you get at this point is not as crucial as simply choosing one and using it as directed (see "Sharpening" on the facing page).

The best sharpening abrasive for beginning woodworkers is fine-grit wet or dry sandpaper glued to a piece of glass. Buy the sandpaper at an automotive-paint supply house in a range of grits—220, 320, 600, 1,000, 1,500, and 2,000. You can find spray adhesive at art supply stores.

turn. Keep your tools sharp, and you'll be surprised at how quickly you become good at woodworking.

How often should you sharpen your edge tools? It depends on the job at hand, but the best indicator is your grumpiness level. When the tool is sharp, the work is satisfying and fun. A dull tool requires more effort and it's not as easy to control. You'll start making mistakes, and wood that planed easily will begin to tear out. When you start thinking this isn't fun anymore, it's time to sharpen (usually after about 15 minutes of continuous work). Better yet, touch up the blade before you get grumpy.

SHARPENING JIG

A good jig ensures a sharp edge every time.

Sharp tools make woodworking almost effort-less, and sharpening is a task you must learn. The first step is to flatten the back of the plane or chisel blade. Lay the blade back-side down on the coarsest sandpaper in your sharpening kit (see the photo at top left), and move it back and forth until the whole surface is uniformly dull. Don't be surprised if this takes more than half an hour. Progress through your sandpapers from coarse to fine, spending three or four minutes on each until the blade's surface is shiny.

Then put the blade in your sharpening jig set to 30° and focus on the bevel, running it back and forth over your second-finest grit for about a minute (see the photo at top right). Once you can feel a little burr across the full width on the back side of the blade, go to your finest grit and hone for a minute or so. Finally, remove the burr by working the back a few times over the finest grit of sandpaper.

Check the edge by holding the blade lightly in one hand and letting it touch the thumbnail on the other hand. A sharp edge will catch on the nail. A dull one will skate over the surface.

You need only hone the very tip of the bevel.

Clamps

Here's a tip that will profoundly affect your woodworking life: Learn to use clamps. Most new woodworkers think clamps are just for holding stuff together while glue dries, but clamps are much more than that. They should be your main method for stabilizing your work. To prevent misaligned joinery, boards shifting as you work on them, and miscellaneous damage from dropping or bumping your work, use clamps.

What to Buy

You'll need a varied clamp collection to meet common shop clamping situations. C-clamps are simple, strong, and inexpensive, and they work well in tight spots. Longer bar clamps are fast, versatile, and good for jobs requiring many clamps. Pipe clamps are indispensable when you need clamping power over long lengths. If your budget allows, you can supplement them with aluminum panel clamps for easy use in light-duty situations. For quick work that doesn't require extreme strength, Quick-Grip® clamps can serve as a convenient extra hand. And though handscrews require two hands to tighten, they are great for holding work that isn't parallel and are just about the only clamps you can clamp down to the bench.

WHAT CLAMPS CAN DO

■ JOIN BOARDS

Panel clamps help keep glue-ups flat.

■ PROVIDE A THIRD HAND

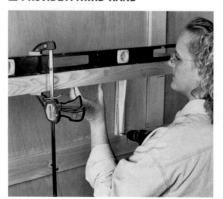

Clamps hold parts and tools steady.

■ GLUE UP

Square up a frame with proper clamping.

Basic Clamps

Start your clamp collection with a few well-chosen types that will serve most purposes and plan to buy more as needed for specific jobs. Having a few of each type is a good start.

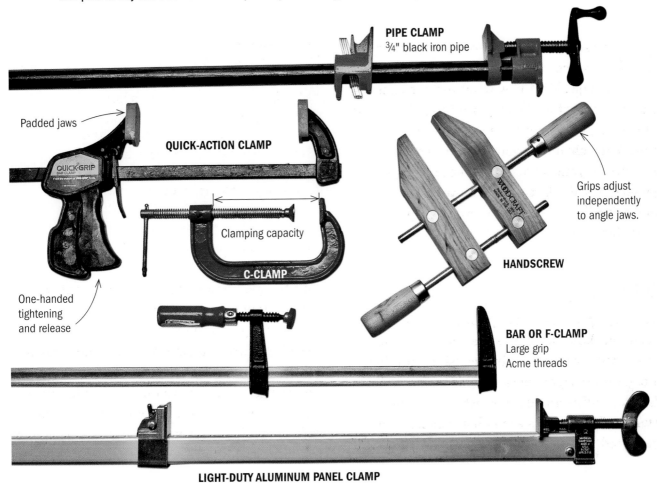

PIPE CLAMP
¾" black iron pipe

Padded jaws

QUICK-ACTION CLAMP

HANDSCREW

Grips adjust
independently
to angle jaws.

Clamping capacity

C-CLAMP

One-handed
tightening
and release

BAR OR F-CLAMP
Large grip
Acme threads

LIGHT-DUTY ALUMINUM PANEL CLAMP

■ **LAMINATE CURVES**

Shopmade jigs allow you to glue curves.

    When buying clamps, look closely at the grip and the threads. A small grip is difficult to turn under high clamping loads and may not achieve adequate pressure. Look for a large, comfortable handle and wide, flat-looking Acme threads, which are easier to wind and unwind. No matter how many clamps you buy, you'll still have jobs that use every one you own.

Sanders and Shaping Tools

When joints don't line up the way they should, new woodworkers often grab a sander and lean into the high spot. Peering through the resulting dust cloud, it may seem that things are getting better, but they probably aren't. Sure, the transition at the joint is smooth, but it's not flat—it sits in a hollow dug by the sander. A shiny finish will magnify the flaw.

It's a classic problem that occurs when sanding by hand or machine. You can't get a flat surface when sanding. Sanding is for smoothing surfaces that have already been planed flat and for preserving that flatness by using diligence and the proper technique.

The first tenet of careful sanding tells us to keep flatness in mind. If you're working on a flat surface, keep it that way. Start by applying only light pressure using a coarse abrasive (80 grit or 100 grit for hardwoods, 120 grit for softwoods) on a stiff pad or block—cork or felt blocks are the traditional favorites, but rubber works well, too. Don't linger in any one spot

and work the entire surface until it's uniformly scratched. If the surface isn't flat and you have to remove high spots, keep a random motion until the high spot is level to the rest of the surface.

The second tenet of careful sanding says that smoothness is best accomplished by progressing from coarse to fine grits, carefully cleaning the surface to remove the previous grit before progressing. If you're using a machine, cover the entire surface equally and finish off with a few minutes of hand-sanding in the

WHAT THESE TOOLS CAN DO

■ SAND FOR FINISHING

Apply light pressure and cover the surface uniformly.

■ POLISH FINISHES, BOATS, AND CARS

6" sanders handle wool bonnets for polishing.

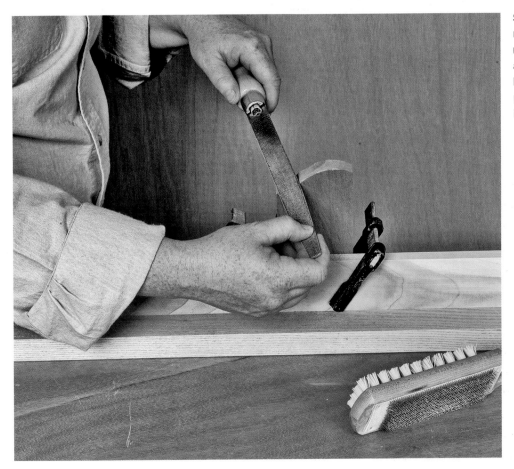

Similar to a file, but with rows of tiny teeth, a pattern-maker's rasp cuts quickly and smoothly. Hold it in both hands as shown and either push or pull it, but don't use it to cut on both strokes.

grain direction using your finest grit and a block. Sand edges by hand only—sanding an edge with a machine is guaranteed to round it over randomly, marring the crisp line for good.

What to Buy

For machine sanding, get a random-orbit sander. The Spirograph® swirl of the sanding disk results from two separate motions—the pad spins in a circle on an eccentric shaft while an

■ SMOOTH A FLAT SURFACE

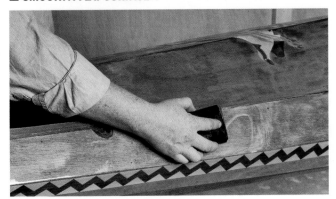

A sanding block helps retain flatness.

■ CONTOUR WITH FORMS

Profiled rubber sanding forms aid sanding in tight spots.

Sanding and Shaping Tools

For efficient machine sanding, use a random-orbit sander. You'll also need hand-sanding blocks and a rasp for fast contouring. Use a file to smooth surfaces after rasping, but before sanding.

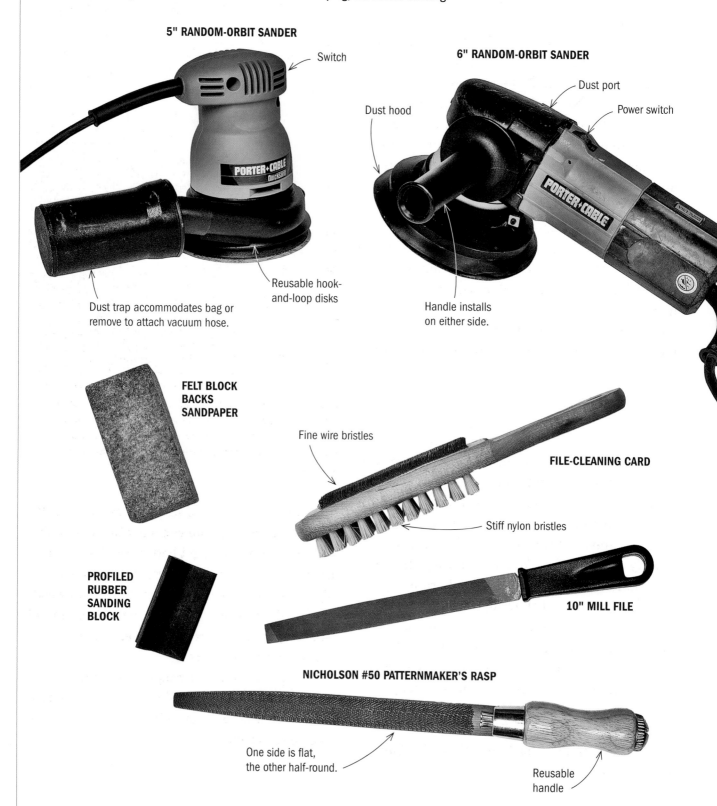

5" RANDOM-ORBIT SANDER

Switch

Dust trap accommodates bag or remove to attach vacuum hose.

Reusable hook-and-loop disks

6" RANDOM-ORBIT SANDER

Dust port

Dust hood

Power switch

Handle installs on either side.

FELT BLOCK BACKS SANDPAPER

Fine wire bristles

FILE-CLEANING CARD

Stiff nylon bristles

PROFILED RUBBER SANDING BLOCK

10" MILL FILE

NICHOLSON #50 PATTERNMAKER'S RASP

One side is flat, the other half-round.

Reusable handle

offset weight spins a slow elliptical orbit. The action isn't truly random, but if you keep the sander moving, it amounts to the same thing. That means two important advantages over other sanders: It leaves few scratches behind, even when sanding against the grain; and it cuts aggressively without being hard to control.

Sanders with 5"- or 6"-diameter pads sand equally well. The smaller sander is handy inside cabinets and other confined spaces. The bigger sander is a two-handed machine, more comfortable for big jobs. Whichever you choose, look for a variable-speed machine with hook-and-loop pads, because they can be reused. A dust-collection port is preferable to a dust bag, and a dust hood—a close-fitting, flexible rubber hood between the sander and the surface—helps prevent airborne dust. Also look for a machine that has a pad dampener to slow the motor when it's not actually sanding. With it, you can lift a running sander to check the surface and replace it without gouging the wood.

Choosing and Using a Rasp

A rasp is an interesting shaping tool that excels at shaping contours and flattening small areas. It's similar to a file, but designed for cutting wood instead of metal. The best rasps have irregular rows of sharp little teeth that cut quickly and smoothly. You'll use a rasp for jobs like rounding corners, shaping end grain, fitting tight joints, and shaping curves.

A trip to a hardware store or home center could turn up five or six different rasps, all of them too coarse for fine woodworking. The only rasp worth buying is a patternmaker's rasp such as the Nicholson® #50. With its close, fine teeth, it's easy to control. Rasps are usually sold without handles; pick up a threaded handle you can reuse.

Random-orbit sanding with dust extraction. A random orbit sander is your best bet for an all-around sander. It's moderately aggressive but doesn't leave swirl marks. For best results, hook it up to your shop vacuum—it'll run cooler and sandpaper will last longer, not to mention keep the air clean.

Even a fine-toothed rasp leaves some pretty rough tooth marks. You can smooth them with sandpaper, but you'll get better results if you use a smooth-cut mill file first. Though it's really a metalworking tool, it works well on wood.

You'll also want to use the mill file on metal —notably for smoothing the edges of planes and preparing scrapers.

Keep your files and rasps clean with a two-sided brush called a file card. It has aggressive metal bristles on one side, and softer nylon ones on the other.

Safety Gear

Hazards abound in a woodworking shop, from the obvious to the insidious. To stay safe and healthy, you must pay attention. As a teacher of mine used to say, "Eternal vigilance is the price of safety."

Eyes

With all the dust, chips, and splinters that machine woodworking sends flying, it's not surprising that woodworkers account for thousands of eye injuries each year. Nearly all could have been prevented had the woodworkers worn protective lenses.

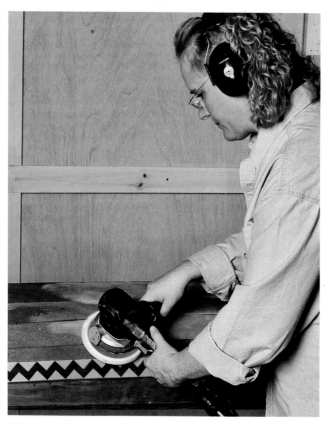

With good dust collection, **you can forego a dust mask and special eye protection when using some hand power tools.**

Find something that fits right and looks good so there's no excuse not to wear it. Look for polycarbonate lenses that meet the ANSI Z87.140 specification (it's displayed on the packaging). If you wear prescription eyeglasses, their lenses probably aren't designed to withstand much frontal impact, and they don't have side shields. You can clip on side shields or buy goggles to fit over your glasses, but the safest solution is to visit your optician for proper prescription safety glasses.

Ears

Noise damage to your hearing is cumulative and irreversible; woodworking machinery is noisy enough to cause damage. Ergo, you need hearing protection. For intermittent use, earmuffs are your best choice. For long-term use, foam plugs are more comfortable and offer more protection. For the best protection, wear both.

Respiratory System

The dust in shop air is the thing most likely to cause health problems. Woodworkers commonly suffer from asthma, sinusitis, bronchitis, and shortness of breath, and they are more likely to contract a rare form of nasal cancer.

It doesn't take much dust to cause problems. The American Conference of Governmental Industrial Hygienists recommends that workers be exposed to no more than a teaspoon of dust in a two-car garage.

Personal Safety Equipment

Working smart and keeping a tidy shop go a long way toward keeping a workplace safe, but some accidents are beyond your control. Guard against surprises by keeping and using all the proper safety gear.

Goggles fit over glasses or can be worn alone.

Seal keeps out dust

Earmuffs

Safety glasses

Foam earplugs

Side shields

Mask for wood dust

COATED COTTON
Nonslip gloves

NITRILE
Durable, disposable, nonallergenic, good for most shop solvents

LATEX
Not good for oil-based solvents

HIGH-TECH CLOTH
Reinforced with Kevlar® at stress points, general-purpose gloves

LEATHER
General-purpose, heavy-duty gloves

The dust you can see in your shop is only part of the problem. The most troublesome dust is the particles that are smaller than 10 microns, and they're invisible to the naked eye. They bypass most of the body's defenses and can enter directly into the sinuses and lungs. They also remain in the air for hours, so be sure to keep your mask on even after the air seems clear.

Disposable masks are fine for dealing with this dust, but be sure to get one labeled for wood dust. You'll get a better fit if it has two bands around the head and a metal strip over the bridge of the nose. Pinch the metal close to your nose to prevent your glasses from fogging.

Hands

Your hands are subject to a variety of dangers when woodworking—sharp tools, splinters, vibration, chemicals, and dirt. Protect them with the right gloves for the job. You'll need a few different kinds:

Leather

Leather all-purpose gloves protect against splinters and blistering. Cowhide is cheaper, deerskin more supple.

High-tech cloth

With spandex backs and Kevlar stress points, these lightweight gloves provide comfort and dexterity for just about any job.

Coated cotton

Close fitting and comfortable, these gloves are good for handling sheet goods and sur-faced lumber.

Vinyl

Vinyl gloves are readily available and inexpen-sive, but they tear easily and offer little protec-tion against most solvents.

Latex

Disposable latex gloves are good for working with epoxies, glues, and water-based dye finishes, but they don't stand up well to oil-based solvents.

Nitrile

Nitrile gloves provide the best disposable pro-tection from the solvents found in woodworking shops, with the exception of lacquer thinner. They offer good dexterity and more durability, and they are nonallergenic.

SHOP LOCATION: GARAGES VS. BASEMENTS		
	Garages	**Basements**
Pros	■ Good access through overhead doors ■ Sturdy floors ■ At-grade access makes it easy to move in machines	■ Electrical service usually nearby ■ Warmth provided by furnace ■ Sturdy floor ■ Overhead storage in floor joists
Cons	■ May need heat or ventilation added ■ May need additional power supply	■ Potential moisture problems ■ Low overhead space ■ Wood dust can infiltrate the house ■ Below-grade access could make it difficult to move in machinery ■ Close location to furnace a potential fire hazard for dust and chemicals

The Essential Shop Space

You'll have to create your first shop wherever you can, most likely in the garage or basement. At this point, you're not ready to devote much space, time, or money to the project, but it's important that you carve out an area dedicated to woodworking (see the floor plan below). If getting at your bench requires much rearranging or means you have to unfold some clever contraption, you'll soon decide it's not worth the hassle. If you can set aside even a small place that's just for woodworking, you'll be able to go out and work for an hour or two without going through a big production.

Set up your bench and clear some space around it to store your tools, materials, and works in progress. The ideal spot has good access for getting in materials and space for working on big projects—even if you have to shift things around occasionally. You'll also need good ventilation and plenty of light.

Wherever you choose, command that space. Don't make do by merely shoving things aside or working on trash cans or on the floor. Clear out your space and stow things neatly. When you're doing woodworking, stop periodically and put away the tools you're not using at that moment. Take the time to use clamps, sawhorses, and other aides rather than pretending you'll get by without doing it right. Develop these habits from the start, and you'll generate fewer mistakes and have more fun—in both the short and long runs.

Floor Plan, the Essential Shop

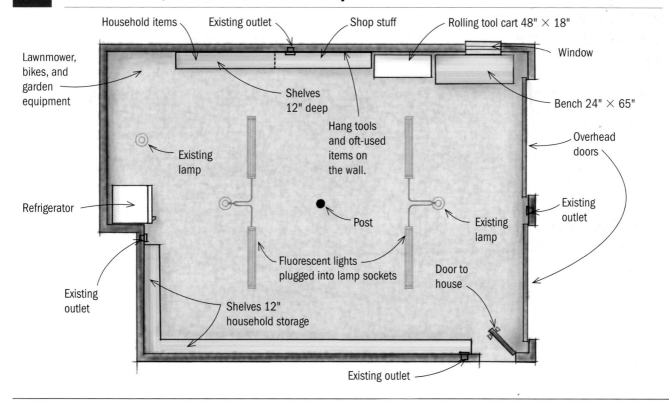

Household items — Existing outlet — Shop stuff — Rolling tool cart 48" × 18"

Lawnmower, bikes, and garden equipment

Window

Shelves 12" deep

Bench 24" × 65"

Hang tools and oft-used items on the wall.

Existing lamp

Overhead doors

Refrigerator

Post

Existing lamp

Existing outlet

Fluorescent lights plugged into lamp sockets

Door to house

Existing outlet

Shelves 12" household storage

Existing outlet

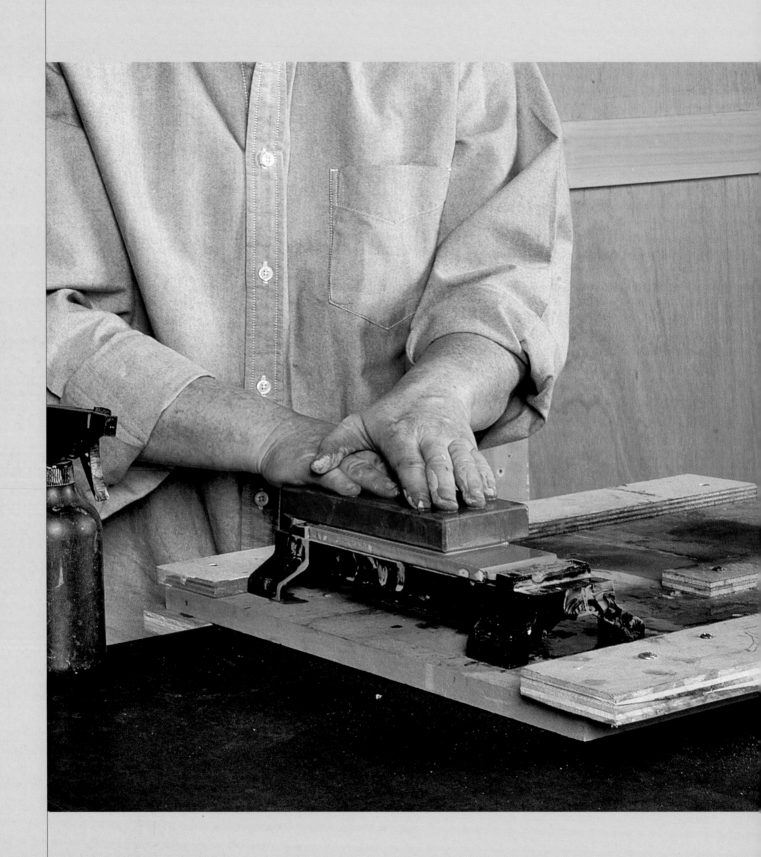

The Basic Shop

As you gain experience with the tools in your First Shop, you'll want to take on more complex projects. Maybe you want to build a bookcase with graceful arches or a sturdy end table with mortise-and-tenon joinery. Perhaps you envision a jewel-like cabinet with dovetailed drawers and hand-carved pulls. The tools in this section make such things possible.

What to Consider

Keep it basic. **There may be faster ways to do things than by using the tools in this section, but the result won't necessarily be better. Basic tools well handled can produce some very fine work.**

To the First Shop's breakdown tool—the circular saw—we can add the miter saw for fast and accurate crosscuts and the jigsaw for cutting curves. With a miter saw, you can take surfaced lumber and break it down into workpieces of nearly any size or shape.

Adding a router to your shop gives you the ability to do all kinds of interesting things to those workpieces. You can cut rabbets, dadoes, and grooves. You can contour edges, mold interesting profiles, and create inlays and other decorative effects. With the right jigs, you can use your router to cut an array of joints, from simple shoulder joints to dovetails and mortise and tenons.

The hand tools in this section allow you to refine the machine work so joints fit better and the finish is finer. And if you're inclined to put in the practice time, you can fashion the same joints with hand tools. The tools in this section are all you need for even the most complex joints.

Hand-tool joinery requires a proper woodworker's bench with a vise or two and a row of dog holes. A bench like this makes working

Keep it organized. You'll need more storage and bench space in your shop to accommodate your new level of skill.

on complex pieces a pleasure, especially if you need to clamp and unclamp frequently to work on a subassembly or to check a joint's fit.

More Tools Need More Space

Your shop will grow to accommodate your higher level of woodworking skill. You'll need more storage space for lumber and tools and more horizontal space to do your work. It takes some thought to organize the growing complex-

ity, but no great financial investment to have a well-run shop.

You'll be surprised at all you can accomplish with these simple tools. Consider the stunningly complex and perfectly finished furniture of the 17th and 18th centuries (think of Louis XIV's palace at Versailles and Chippendale). The lion's share of a master cabinetmaker's tools from that era were nothing more than customized measuring and marking tools, planes, chisels, and handsaws.

Basic tools, yes. The results can be anything but.

Router

Given a little ingenuity and a good jig, you'll find there's not much a router can't do around the shop. I've seen them used for everything from making dowels and roughing out backgrounds of carvings to profiling elegant edges and cutting intricate joints.

The router is a very simple tool, and therein lies its versatility. It's a motor turning at some 27,000 rpm connected to a bit with a tapered collet that squeezes the bit's shank when a special nut is threaded tight. This unit locks into a handy base that keeps it upright and allows for adjusting the depth of cut. This simple arrangement accepts a multitude of router bits in all shapes and sizes—from 3/16" straight bits to complex bits with diameters of 1½" or more.

Whole books have been written on the subject of routers, and you should study them. Before long, you'll be creating router jigs to solve your own woodworking problems.

What to Buy

Start out with a handy midsize router (1½ hp to about 2 hp) with collets to fit bits with either ½" or ¼" shanks (see "What Size Shank?" on p. 35). For maximum versatility, get a kit that includes interchangeable fixed and plunge bases (see "Router with Fixed and Plunge Bases" on p. 32). You'll use the fixed base most of the time

WHAT A ROUTER CAN DO

■ JOINT

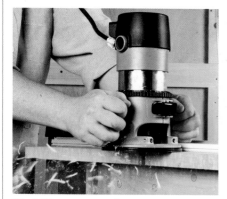

Straighten a wavy edge.

■ PLUNGE

Lower a spinning cutter into a workpiece.

■ PROFILE

Cut a decorative pattern rabbet in an edge.

Approach is key. Control the router at all times. Develop a strong stance and get down low enough so you can really see what's going on.

because it's lighter and better balanced, but the plunge base lets you safely lower the router bit into the work for routing in the middle of a workpiece. Look for a special sale on a kit that includes both bases—buying bases separately is considerably more expensive.

Your next criteria should be comfort. Find a router that feels right in your hands, one on which the knobs are the right size for you, the on/off switch is conveniently located, and the depth adjustment is easy for you to operate. Check out the plunge base as well and make sure you can operate the plunge-release lever

■ **DADO**

Cut a groove across the width.

■ **CUT TEMPLATES**

Cut an exact copy of a template.

Router with Fixed and Plunge Bases

A router in the 1½-hp to 2-hp range is big enough to do most jobs and not too unwieldy. Be sure to get a router with interchangeable fixed and plunge bases. An optional D-handle matched with an offset base gives a wide footprint ideal for routing edges.

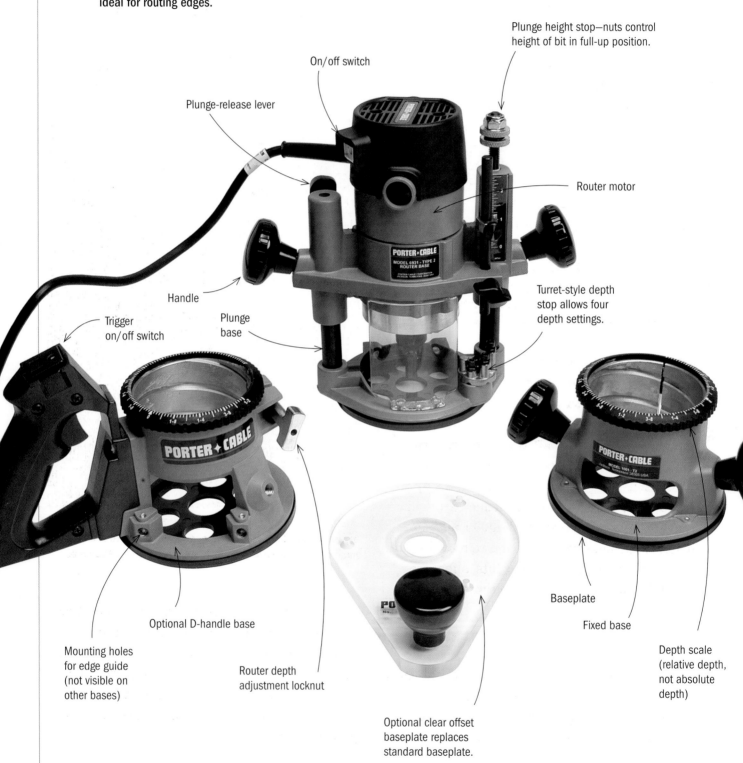

Plunge height stop—nuts control height of bit in full-up position.

On/off switch

Plunge-release lever

Router motor

Handle

Turret-style depth stop allows four depth settings.

Trigger on/off switch

Plunge base

Mounting holes for edge guide (not visible on other bases)

Optional D-handle base

Router depth adjustment locknut

Optional clear offset baseplate replaces standard baseplate.

Baseplate

Fixed base

Depth scale (relative depth, not absolute depth)

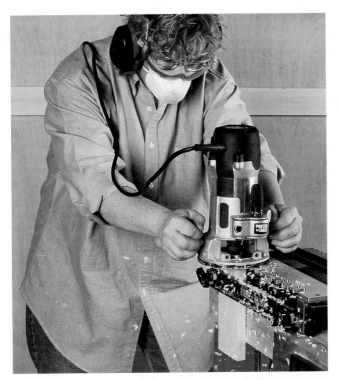

Versatile jigs. You can do almost anything with a router and the right jig. This one routs both parts of a dovetail joint at once. Once it's set up correctly, you can easily rout a kitchen full of identical dovetailed drawers.

easily. And take a look at the depth stop—you'll want one that adjusts to cut at accurate depths, without needless complexity.

Then you can consider the niceties. A spindle lock button holds the arbor from turning so you'll need only one wrench to change bits. A soft-start motor gradually ramps up to speed, eliminating the high-torque twitch of a router going from 0 rpm to 27,000 rpm in an instant. Variable speed lets you reduce the rpm when turning large-diameter bits, a feature you'll find handy from time to time.

Beyond Edges

When you first start using a router, you'll mostly use it for dealing with edges—straightening, profiling, or making them more interesting (see "Five Ways to Guide a Router" on p. 34). These

Inlaid detail. You can create perfectly fitting inlays of any shape with a router, a template, and the right collars. The inlay will fit perfectly, with just enough space for the glue.

everyday tasks are a great way to get familiar with the router, but you'll soon find other uses. One of my favorite router applications is making identical parts from a template. Once you have a good template, it's easy to run a straight bit with a bearing around the edge for perfect copies.

SHANK-MOUNTED BEARING

A template secured atop the workpiece guides a bearing mounted on the bit's shaft. If the bearing's diameter is equal to the bit diameter, it produces an exact copy of the template.

TIP-MOUNTED BEARING

Tip-mounted bearings usually run against the workpiece's edge or against a template. They're often used for molding decorative profiles, for cutting rabbets, or for use with templates mounted below the workpiece.

COLLAR

A collar screwed in the base plate runs against a template or guide, offsetting the cut from the edge. There's some math involved to position the cut correctly, but collars offer a wider depth adjustment range than do bits with bearings.

FENCE

Some router baseplates feature a straight side to run along the fence. If yours doesn't, put a mark on your base and keep that point against the fence.

EDGE GUIDE

A fence attached to the baseplate makes simple work of routing lines parallel to edges. Modify the guide for less contact surface, and it will follow a curved edge.

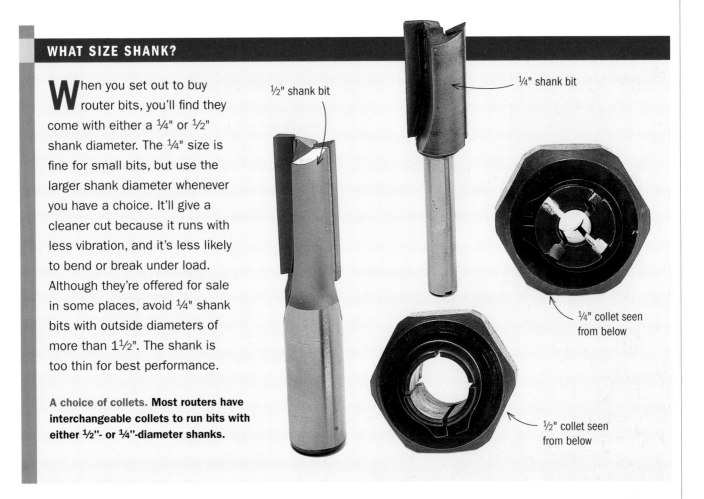

½" shank bit

¼" shank bit

¼" collet seen from below

½" collet seen from below

When you set out to buy router bits, you'll find they come with either a ¼" or ½" shank diameter. The ¼" size is fine for small bits, but use the larger shank diameter whenever you have a choice. It'll give a cleaner cut because it runs with less vibration, and it's less likely to bend or break under load. Although they're offered for sale in some places, avoid ¼" shank bits with outside diameters of more than 1½". The shank is too thin for best performance.

A choice of collets. Most routers have interchangeable collets to run bits with either ½"- or ¼"-diameter shanks.

With the right jigs and bits, routers can solve a multitude of problems. They can cut perfectly fitting inlays—either thin ones for decoration or thicker ones to fill a knothole or make a structural repair. Mounted on an arm pivoting around the center point, the router is an accurate and secure way to cut a perfect circle. A router can cut the classic joints of hand woodworking—the mortise and tenon and the dovetail (see the top left photo on p. 33)—and it can straighten and smooth as well.

Perfect circles every time. The router is a great way to cut circles. For perfect arcs, mount it on a long arm and pivot it around a nail or screw.

Jigsaw

Using a jigsaw adds a new dimension to your woodworking—curves. From the long arches of an Arts and Crafts style bookcase to the tight scrolls of a decorative shelf bracket, this saw can handle every type of curve. It also handles straight cuts, and it is one of the safest tools to grab for making quick cuts by eye. Because the blade on a jigsaw moves up and down, as long as you keep the sole on the workpiece and don't twist the blade too much, it won't kick back or otherwise misbehave.

Making a cut with a jigsaw is easy. If the saw is hard to push or steer, something's wrong. The blade may be the wrong type or dull, the cutting speed may be wrong, or the orbital adjustment may be set incorrectly for the type of cut. The orbital adjustment adds a back-and-forth component to the saw's usual up-and-down motion.

The highest orbital setting cuts aggressively, powering through the wood and leaving a ragged edge. Use this setting for rough cuts in solid wood. For smooth cuts, zero out the orbital action. The saw cuts more slowly, but leaves a smoother edge.

Most jigsaws have variable speed. Some build it into a sensitive trigger switch; others use a dial for speed control and a simple on/off switch. In most cases, you won't need to vary the speed as you cut, but you might need to vary the speed to suit the material you're cutting. Use lower speeds for metals, higher speeds for wood.

What to Buy

Don't bother buying a jigsaw without orbital action; it's not up to serious work. Make sure you choose one with electronic speed control, and a base that tilts in both directions. Have a good look at the blade lock and choose a method that's simple and secure. Some of the "tool-less" methods are so fussy it's easier to use a tool.

WHAT A JIGSAW CAN DO

■ CUT CURVES

Cut freehand curves from flowing to tight.

■ CUT ANGLES

Tilt the blade to the right or left.

■ CUT SCROLLS

Start your cuts in a drilled hole.

Orbital Jigsaw and Jigsaw Blades for Woodworking

Look for a machine that has adjustable orbital action and variable speed. Triggers that lock easily into the "on" position are useful, and a tilting base allows you to make cuts at angles other than 90°. The right blade is crucial for best performance. Change blades often—that's why they come in packs of five.

Blade lock (this saw uses a screwdriver to turn a screw)

Button locks trigger "on."

Allen wrench for tilting base

Electronic speed control

Slots for fitting edge guide

Variable-speed trigger

Air blower adjustment for clearing sawdust

Blade guide

Orbital adjustment

Not shown—plastic shoe for sawing materials that scratch easily

■ CUT STRAIGHT

Use a straightedge or guide.

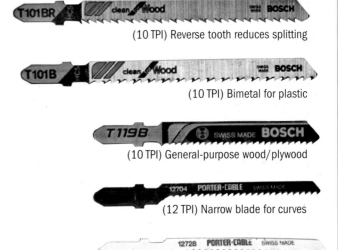

(6 TPI) Use for rough cuts

T101BR (10 TPI) Reverse tooth reduces splitting

T101B (10 TPI) Bimetal for plastic

T119B (10 TPI) General-purpose wood/plywood

(12 TPI) Narrow blade for curves

(12 TPI) Cuts curves in plywood

Bench Planes

Many modern woodworkers ignore the bench plane under the illusion that power-driven tools are somehow better. Compared to a plane, these crude tools don't do most jobs as well or as easily. Once you know your planes, it's a snap to set one to remove superfine shavings—as thin as four ten-thousandths of an inch thick—to sneak up on the fit of a joint or to smooth rough swirls of grain.

What to Buy

First get a 9" #4 or #4½ smoothing plane. I like the wider #4½ because it uses the same irons as the #7. This smaller plane is best for smoothing surfaces after they're flat and for flattening smaller workpieces.

Then get a #7 jointer plane, about 22" long. For flattening, it's the best tool in your shop. You should use it whenever the workpiece is large enough to support it.

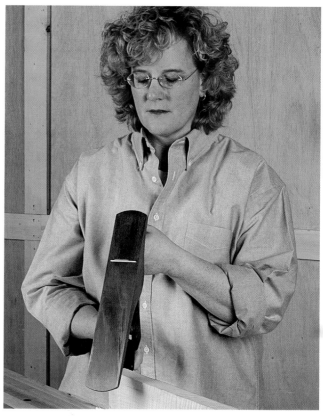

Tuning up. To get top results from a plane, you'll have to spend some time practicing with it, getting to know its ways, and making sure it's sharp and properly adjusted.

WHAT BENCH PLANES CAN DO

■ JOINT

Flatten and square an edge.

■ SMOOTH

Make a surface smooth while maintaining flatness.

■ FLATTEN

Remove bumps and hollows to make a surface level.

Bench Planes

These three planes will handle most of your planing jobs, from rough to supersmooth finishing.

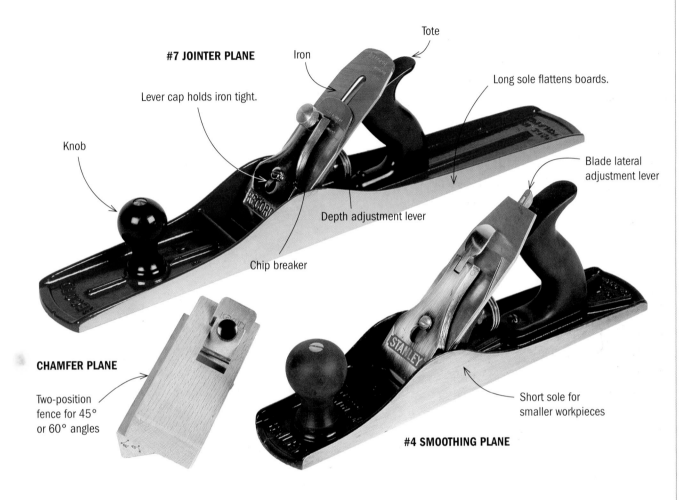

#7 JOINTER PLANE

Tote

Iron

Lever cap holds iron tight.

Knob

Long sole flattens boards.

Blade lateral adjustment lever

Depth adjustment lever

Chip breaker

CHAMFER PLANE

Two-position fence for 45° or 60° angles

Short sole for smaller workpieces

#4 SMOOTHING PLANE

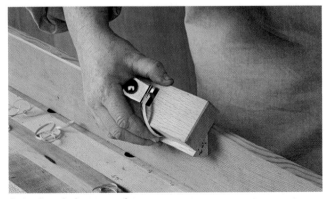

■ CHAMFER

Cut a bevel along an edge.

You'll have to look a little to find these planes—most places sell only the #5 jack plane, supposedly the "jack of all trades." It's really too short for jointing and too long for anything else. It's better to start with two planes and let each do the job it does best.

You can do all the chamfering you need with these two planes and your block plane, but a dedicated chamfer plane ensures that your chamfers are all at the same angle and depth, something you'll notice on long edges.

Sharpening Tools

It's been said that civilization began with learning to sharpen; improving the odds of the hunt led to good health and allowed enough free time to develop art. You can expect similar results in your woodworking. Until you get sharpening down cold, you'll be struggling too much to have fun or get beautiful results.

The sandpaper sharpening method described on p. 15 is a great way to get started sharpening, but when you begin building more complex pieces, you'll need the best edges you can get (see "Five Steps to a Perfect Edge" on p. 43). Sandpaper is too coarse to produce the finest edge.

What to Buy

The best way to flatten the backs of edge tools is with a coarse (about 220-grit) stone. I prefer diamond stones because they're so hard they cut the fastest. Plus, diamond stones don't dish with use and stay so flat you can use them for

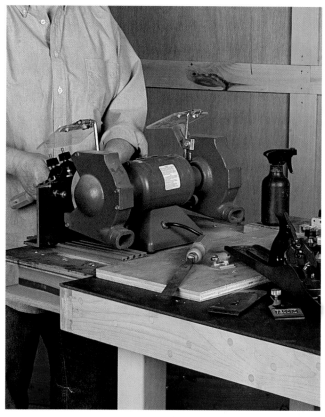

A task worth the time. Truly sharp tools require investments of your time and money. The payoff is high—better results, less effort, and a lot more fun.

WHAT THESE TOOLS CAN DO

■ FLATTEN BACKS

Flatten the backs of planes, chisels, and other cutting tools.

■ GRIND

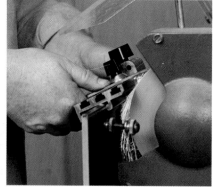

Rough out the bevel at 25° for faster honing.

■ HONE

Refine the edge at 30° to sharpness.

Tools for Flattening and Honing

It takes a lot of gear to get the sharpest edge, but as always, the right tools make all the difference.

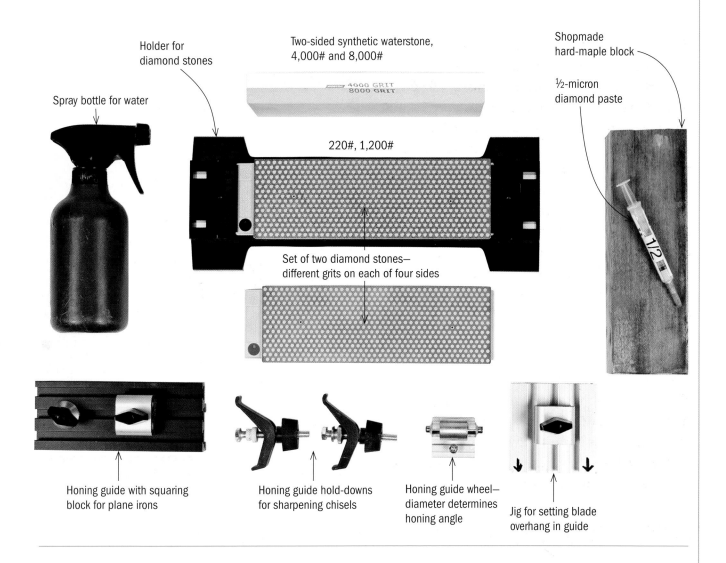

Holder for diamond stones

Two-sided synthetic waterstone, 4,000# and 8,000#

Shopmade hard-maple block

½-micron diamond paste

Spray bottle for water

4000 GRIT 8000 GRIT

220#, 1,200#

Set of two diamond stones— different grits on each of four sides

Honing guide with squaring block for plane irons

Honing guide hold-downs for sharpening chisels

Honing guide wheel— diameter determines honing angle

Jig for setting blade overhang in guide

flattening other types of stones, plane soles, and more. Whatever type of stone you get, you'll need three or four stones ranging from 220 to 1,200 grit.

Your first honing guide might not be wide enough to accommodate a #7 plane iron—now's the time to upgrade. You'll also need a guide for your grinder. You can buy the two items individually or get a modular system that uses the same parts for both.

Get a slow-speed grinder (1,800 rpm or so) and a soft wheel for sharpening. This combination runs cooler and reduces the chances of ruining the blade's temper from overheating.

Synthetic waterstones give the best edge, and you'll need at least two grits. Depending on the brand, the coarser grit should be 3,000# to 4,000# and the finer 6,000# to 8,000#. Get a single two-sided stone, or two separate stones.

The Slow-Speed Grinder

A slow-turning grinder with a soft wheel runs cooler and keeps your blades from overheating and losing temper. Either a 6" or 8" wheel is fine, though the 6" wheel has a lower velocity at the rim.

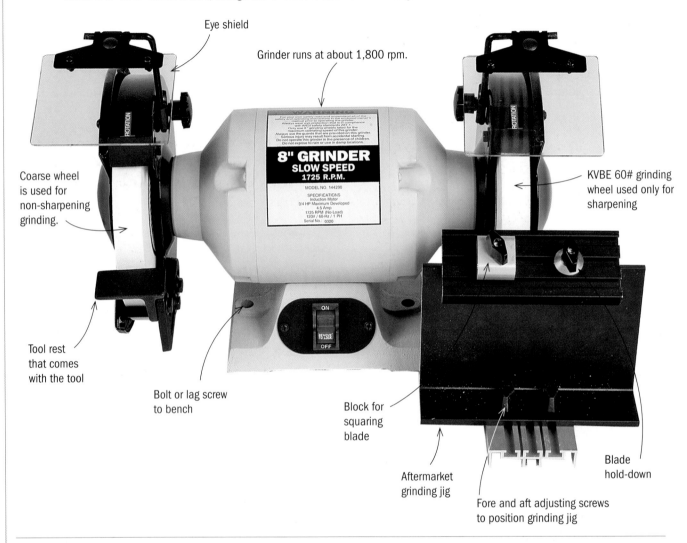

Eye shield

Grinder runs at about 1,800 rpm.

8" GRINDER
SLOW SPEED
1725 R.P.M.

MODEL NO. 144290

SPECIFICATIONS
Induction Motor
3/4 HP Maximum Developed
4.5 Amp
1725 RPM (No-Load)
120V / 60 Hz / 1 PH
Serial No.: 0320

Coarse wheel is used for non-sharpening grinding.

KVBE 60# grinding wheel used only for sharpening

Tool rest that comes with the tool

Bolt or lag screw to bench

Block for squaring blade

Aftermarket grinding jig

Fore and aft adjusting screws to position grinding jig

Blade hold-down

Finally, for the ultimate edge, you'll need some diamond paste and a little mineral oil to cut it. Since diamond paste is usually sold in kits for gem polishing and electronics, you'll probably end up with four or five grits from 9 microns to ½ micron. You only need the finest for honing, but you can use the others to polish the backs of your blades. Use the diamond paste on thick maple blocks cut to the same size as your sharpening stones.

Flat stones are crucial to a good edge. **Flatten them often on your diamond stones.**

FLATTEN AND POLISH THE BACK

A perfect edge starts with flattening the back on a fast-cutting coarse diamond stone and progressing to your finest grit stones. For the ultimate mirrorlike polish, finish with ½-micron diamond paste on a flat, smooth block of hard maple.

GRIND AT 25° FOR FASTER HONING

Grind the bevel to 25°. Later, when you hone at 30°, you'll remove material from only the tip of the tool, which takes just minutes. Grind again after four or five honing sessions.

HONE THE BEVEL AT 30°

Start honing with your second-finest stone for about a minute. Then create a slight crown in the edge to prevent the corners from digging. Press a little harder on each corner for several strokes.

POLISH THE BEVEL

Using your finest stone, polish for about a minute, pressing on alternate edges to maintain the crown. Then work the back for about 15 seconds. For ultimate sharpness, polish both sides with ½-micron diamond paste.

TEST THE EDGE

Hold the blade loosely in one hand and gently touch the edge to a thumbnail. A sharp edge bites in, a dull one slips or scrapes. If the iron doesn't pass this test, spend more time working the back on your finest stone.

Measuring and Marking Tools

When you're first learning woodworking, your hand skills (or lack thereof) are your roadblocks to success. Once you understand how to use the tools, a new roadblock arises: your ability to measure and mark correctly. Accurate layouts are mostly a matter of patience, but good tools play an important role.

Moreover, there's a completely new set of skills you need to learn in order to measure properly—things like how to present a square to an edge so it gives a true reading and how to use a knife to mark a line in a way that won't damage your straightedge or the blade. When you reach this level of woodworking, a little bit off is too much.

Quality tools for quality work. **Good measuring tools are essential for fine joinery. Use them to check your handwork often, and correct small issues before they become larger problems.**

What to Buy

If you're serious about doing good work, keep a couple of sliding squares close by at all times (see "The Indispensable Sliding Square" on p. 46). I'm hard-pressed to say which I use most—(see "The Indispensable Sliding Square" on p. 46) a 4" or a 6" square. Get the one that appeals to you most, then ask for the other for your next birthday. You'll find you use a 12" combination square less often, for larger measurements and for laying out 45° angles.

WHAT THESE TOOLS CAN DO

■ LAY OUT 45°ANGLES

Use the 45° leg of a combination square to mark an angle.

■ MARK CUT LINES

Use a knife to scribe lines for accurate joinery.

■ MARK ANGLES

Use a sliding bevel that adjusts to any angle.

Measuring and Marking Tools

Get good tools and use them wisely to ensure that all your joinery comes out square and accurate.

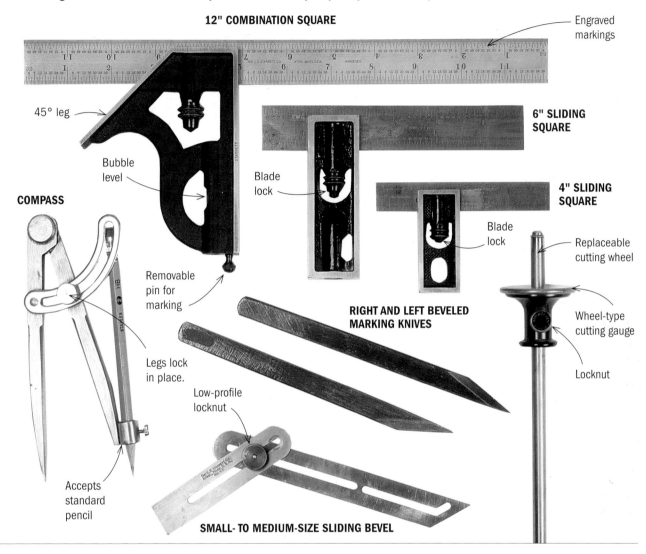

12" COMBINATION SQUARE

Engraved markings

45° leg

Bubble level

COMPASS

Blade lock

6" SLIDING SQUARE

Blade lock

4" SLIDING SQUARE

Replaceable cutting wheel

Removable pin for marking

RIGHT AND LEFT BEVELED MARKING KNIVES

Wheel-type cutting gauge

Locknut

Legs lock in place.

Low-profile locknut

Accepts standard pencil

SMALL- TO MEDIUM-SIZE SLIDING BEVEL

■ SCRIBE A LINE PARALLEL TO AN EDGE

Use a cutting gauge with a fixed knife to scribe a line a set distance from an edge.

■ CHECK SQUARENESS

Make sure the edge is parallel to the side.

TRANSFER LINES

Rather than use a tape measure or ruler to locate a line, put the square on the edge and slide the blade to the line. Lock the blade in place and transfer the measurement without having to read the scale.

SQUARE AROUND A BOARD

Put the point of your pencil on the mark and slide the square up to it. Hold the square firmly and draw the line. Rotate the workpiece and repeat the process right around the board.

CHECK ANGLES

By changing the blade position to just below flush, you can check inside and outside corners for squareness.

MARK A LINE PARALLEL TO AN EDGE

Using the base as a reference point, slide the scale to the desired measurement. Place a pencil firmly against the end of the blade and push the square along one edge.

MEASURE DEPTH

Slide the blade down into the recess and lock it in place. Either read the scale or transfer the measurement directly.

Though used mainly for laying out and marking joints, your measuring tools have other applications. You can use them to check that your hand power tools are set up correctly—for example, you can check the jigsaw's tilt against the sliding bevel and use a square to true up the position of the grinder jig so it's square to the wheel and the right distance from its face.

You should also use the tools for developing your hand skills. Practice with your bench plane on some fragrant pine, using the square to check for perpendicularity at short intervals. Learn how to hold your plane and distribute your weight until the square shows no changes. Set the sliding bevel to a random angle and plane the angle, then plane back to square.

Small- to medium-size sliding bevels (6" or less) are more comfortable to use than large ones. I like the precision of an all-metal tool, but whatever you buy, remember that large locking nuts get in the way. You'll need a pair of marking knives, beveled right and left so you always put the flat back against the straightedge. You'll need a cutting gauge for making knife cuts on face grain and endgrain at a set distance in from the edge of a board. The wheel-type cutting gauge is a real advance from traditional cutting and marking gauges. Finally, seek out a durable compass that accepts a regular pencil and locks the legs in place so they won't shift in use.

A pair of winding sticks that are perfectly straight, square, and true can help you see twist in a board. Sight across the top of the near stick to the top of the far one. If the board is flat, their tops are parallel.

Handsaws

N o matter how many power saws you own, there's always a place for handsaws. With a little practice, they afford a level of control you can't achieve with a power saw. And you'll be surprised how often it's easier to grab a crosscut saw to cut a board to rough length than it is to get out an extension cord and set up a power saw.

What to Buy

Get an aggressive saw for rough cutting to length. Any hardware store or home center offers at least a couple of choices, usually a traditional European-style saw, and a hybrid Japanese-style saw (see "Push or Pull?" on the facing page).

Next, get one or more backsaws—fine-toothed saws with reinforced backs for stiffness. These are tools you'll need for hand-cutting joints like dovetails and mortise and tenons. Your first choice should be one 8" to 10" long,

Success with a handsaw is all about technique. **Align your body to the task and give yourself plenty of room. With a little practice, you'll find that effortless rhythm that means everything's right.**

WHAT HANDSAWS CAN DO

■ ROUGH-CUT LUMBER

A handsaw is faster than a power saw for a cut or two.

■ CUT CURVES

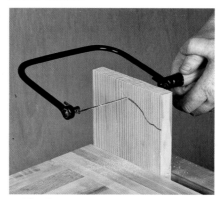

A thin, fine-toothed blade will cut almost any shape.

■ CUT JOINTS

A stiffened blade is best for cutting accurate joints.

Choose Your Style

Start your saw collection with a rough crosscut saw, a backsaw (either European or Japanese style), and a coping saw.

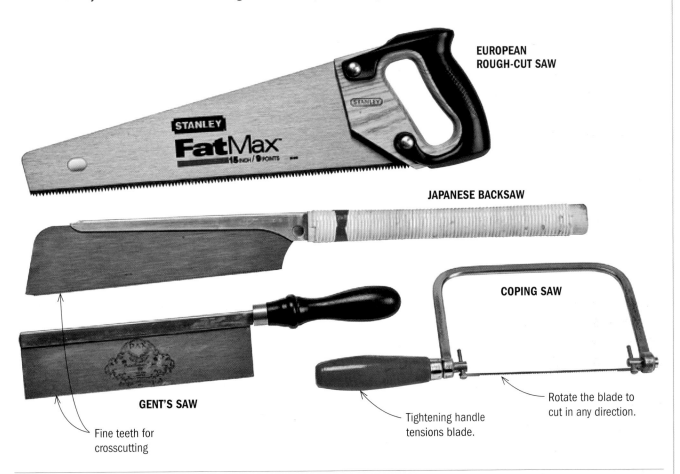

EUROPEAN ROUGH-CUT SAW

STANLEY FatMax 15 INCH / 9 POINTS

JAPANESE BACKSAW

COPING SAW

Rotate the blade to cut in any direction.

GENT'S SAW

Tightening handle tensions blade.

Fine teeth for crosscutting

either European or Japanese style. Within the European style you'll have a choice of the round-handled gent's saw or the more familiar looking dovetail saw. If you make a lot of tenons by hand, you'll also want to get a longer tenoning saw with its unique fine teeth set for ripping.

Then pick up a coping saw. It's the best tool for cutting complex compound curves. One great feature of these saws is that you can rotate the blade to saw at the most advantageous angle. Get a selection of blades—coarse, medium, and fine—to handle whatever job arises.

PUSH OR PULL?

The essential difference between European and Japanese saws is in the teeth. A European backsaw cuts on the push stroke, which is handy for very fine work because the sawdust accumulates at the back and doesn't obscure the view. The teeth on a Japanese backsaw are thinner and cut on the pull stroke. Many beginning woodworkers prefer Japanese saws because they cut fast and clean, and they are easy to control.

Woodworker's Bench

wo things distinguish a woodworker's bench: It has at least one vise, and it has at least one row of dog holes (either square or round) along the front edge. The presence of these two items transforms a table into a big, versatile clamp.

Most traditional woodworker's benches have two vises: one set on the front of the bench near the left side (for right-handers) for most simple clamping operations, and a tail vise on the right end of the bench that's used in connection with the dog holes. A wood or metal pin (the dog) set into one of the dog holes on the left end of a workpiece acts as a stop for planing. Put a dog in the hole on the cheek of the tail vise, and you can crank the vise to squeeze the workpiece between the two dogs for a more secure hold.

The bench shown here makes use of a versatile metal vise on the left end of the bench to fulfill the functions of both a front and tail vise. You can stand at the end of the bench when you

need the vise to hold small objects, and work along the front of the bench when working with long boards or planing.

What to Buy

Whether you build a woodworker's bench or buy one, it should meet the size and height criteria of the Essential Shop workbench (see pp. 6–7). Make sure the rods at the bottom of the jaw are at least 4" below the surface of the bench—if not, you'll find it difficult to clamp wide boards securely. Look for dog holes 4" to 8" in from the edge. Traditional dogs are square, but round ones are more versatile.

Most woodworker's benches have a skirt around the edges; make sure it's at least 1½" thick to provide adequate footing for vertical clamping with big clamps. Look for minimal obstructions beneath the bench so you can clamp across the underside.

Continued on p. 52

WHAT A WOODWORKER'S BENCH CAN DO

■ HOLD PIECES IN A VICE

A vise holds the wood securely for a variety of operations.

■ SECURE WORK WITH DOGS

One dog acts as a stop; a second set in the cheek of a vise secures work.

A Bench That Holds Your Work

Built on the base of the bench used in the Essential Shop, this bench has all the features you need for advanced woodworking. This uncommon design (suitable for right-handers) uses an iron quick-action vise on the left end that functions as both front vise and tail vise.

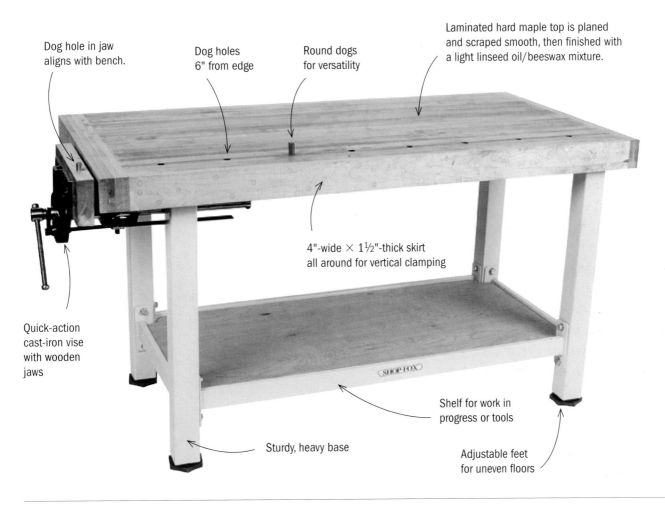

Dog hole in jaw aligns with bench.

Dog holes 6" from edge

Round dogs for versatility

Laminated hard maple top is planed and scraped smooth, then finished with a light linseed oil/beeswax mixture.

4"-wide × 1½"-thick skirt all around for vertical clamping

Quick-action cast-iron vise with wooden jaws

Shelf for work in progress or tools

Sturdy, heavy base

Adjustable feet for uneven floors

■ HOLD CLAMPS

Secure awkward pieces with a combination of vise and clamps.

■ SUPPORT BOARDS FOR PLANING

A wooden L clamped in the vise holds the board in place for easy clamping across the benchtop.

A woodworker's bench is optimized for using hand tools like planes, chisels, and saws. It offers a variety of ways to hold and support the work at hand.

Though in-bench storage seems like a good idea, bench drawers are overrated. They're not really that convenient for holding your tools, and they tend to fill up with dust and shavings. Plus, a few drawers full of tools can add enough weight to a bench to make it difficult for one person to move around the shop. Though you'll store your bench against the wall and often use it in that position, you'll frequently want to pull it out to the center of the shop for access to all four sides. If your floor is uneven, you'll want large, sturdy self-leveling feet.

A light finish seals the benchtop against moisture and makes it easier to clean. An oil-wax or oil-varnish finish works best and is easy to renew.

The Basic Shop Space

Once you commit yourself to woodworking, your relationship to your shop changes. You're spending more time there and building more complex projects, so having to move the bikes, reorganize the garden tools, and set up lights before you can work is no longer tolerable. Your growing tool collection is taking up more room, and there's lumber to store, too (see the floor plan on p. 55).

You'd like more room, but since the available space is fixed, what you really need is better organization.

Cabinets and Shelves

Keep your Essential Shop space intact and use it to build cabinets to replace some of the shelving you're using. Build the cabinets in two banks and cover their tops with ¾" (or thicker) plywood or MDF so the tops are at bench height. Leave room between the banks for your bench, and you'll end up with a whole wall of workspace with surfaces at bench height. You can add more cabinets above the bench, but you'll have better light and more room to work if you don't. Use the space above the bench by installing narrow shelves or hanging frequently used tools there.

Rather than succumb to the desire to build the perfect storage cabinets with complicated features, start by building a batch of simple boxes with doors. Get them in service and figure out what wants to live where. To increase storage space later, you can customize the cabinet to its contents by installing partitions, sliding shelves, and drawers.

Think vertically. Use every bit of wall space—even the ceiling—and don't let the space beneath shelves or tools go to waste. The more you can store, the more room you'll have for woodworking.

Now there's space on the other wall for storing lumber and sheet goods. Store plywood at the back of the shop, leaning the sheets against the wall vertically—or as close to vertical as possible. Store solid wood on well-fastened heavy-duty shelf brackets. Twelve feet of shelving will accommodate lumber up to about 14' in length—longer pieces can rest on 2×4s set on the floor.

Add built-in cabinets and drawers. Hang jigs and infrequently used items high on the walls or even on the ceiling. Use the space between the joists or build low-profile racks for hanging things from the overhead. Use the floor underneath tools, the space beneath the staircase, and the area between the garage doors. Make the most of every inch of space, and your shop can happily coexist with the rest of your things.

When your bench is pushed against the wall with your most commonly used tools hanging nearby, you can get right to work whenever you have a spare moment. But sometimes you'll want to pull it away from the wall to get at all sides of your project.

Floor Plan, the Basic Shop

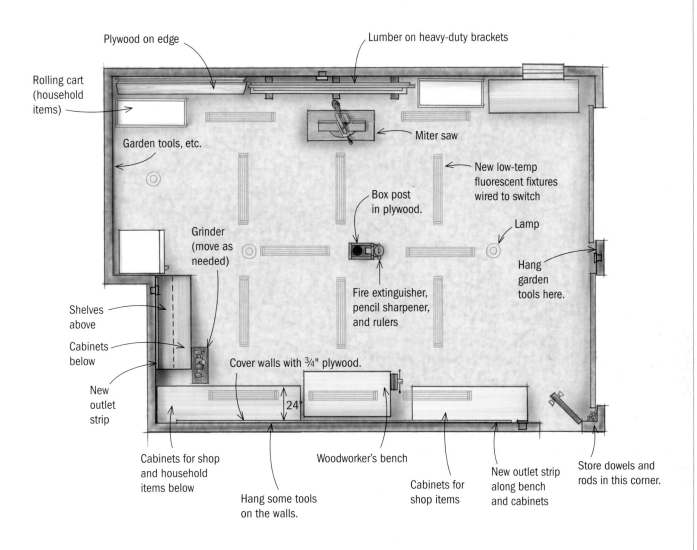

Plywood on edge

Lumber on heavy-duty brackets

Rolling cart (household items)

Garden tools, etc.

Miter saw

New low-temp fluorescent fixtures wired to switch

Box post in plywood.

Lamp

Grinder (move as needed)

Fire extinguisher, pencil sharpener, and rulers

Hang garden tools here.

Shelves above

Cabinets below

New outlet strip

Cover walls with ¾" plywood.

24"

Cabinets for shop and household items below

Hang some tools on the walls.

Woodworker's bench

Cabinets for shop items

New outlet strip along bench and cabinets

Store dowels and rods in this corner.

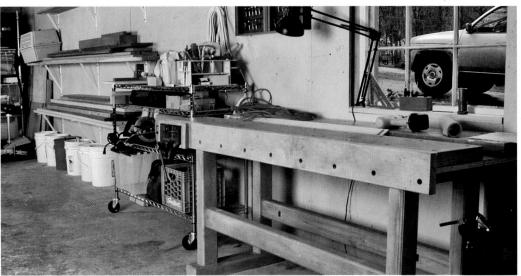

Your first workbench can stay in its original location, and you can reorganize the area around it for storing lumber. The sheet goods are stored in the far corner, accessible by moving the rolling shelves filled with home and garden items.

ROLLING TOOL CABINET

Designed for auto shops, a multitiered metal tool cabinet works just as well in a woodworking shop. It keeps your tools nearby, even when you're not working at your bench. If you're concerned about unauthorized use of your tools, you can lock the cabinet up.

BINS AND CRATES

Moderate-size clear plastic bins with lids are a good way to store stuff in the shop. They're not big enough to get too heavy, and you can see what's in them. Open crates are smaller, more heavily built, and stackable, but small items will slip through the holes in the bottom and sides.

SOFT-SIDED BAGS AND TOOL ROLLS

Tools kept in a soft-sided bag aren't as likely to be damaged as those rattling around in a metal case. With pockets on the outside and divided space on the inside, a couple of moderately sized bags can hold most of your tools. Keep chisels and rasps in a tool roll with pockets.

DOWELS IN THE WALL

Rather than fiddling with pegboard and hangers that always fall out, put your tools on lengths of ½" dowel set into holes drilled in the wall. Don't expect this to work in drywall—cover it with a sheet of ¾" plywood first. Use nails or screws for hanging small items like rulers.

The bank of cabinets shown in this photo is nothing more than a line of identical plywood boxes screwed to the wall. A plywood top spans the bank and matches it to the bench height. Built quickly with a pocket-hole jig, these cabinets filled up almost immediately. MDF doors came later and were dirt-simple to construct—just cut the MDF to size and install the hardware.

Prebuilt cabinets are an even quicker way to add storage, and the cost can be even less than the simplest shop-built cabinets. You'll have to design your space to accommodate the cabinet sizes stocked by your local home center, but you can set up the shop in an afternoon.

Made from ¾" shop-grade plywood with MDF doors, **these cabinets are serviceable and sturdy. You might say they have a rugged and functional handsomeness.**

An array of clamp racks on the wall is an impressive sight, but wall space is at a premium in most shops. When you have only a few clamps, you can keep them in a bucket or bin and push them under the bench when not in use. But clamps in a bin end up in a tangled mess. A better solution is to build a clamp rack on casters. It frees up wall space and puts the clamps close by when you need them. When you don't, just push the rack aside.

Maximize the remaining space by parking some rolling shelves along the back of your shop with little or no space between them. When you need an item from a shelf, simply roll the cart out of line and into the open, much like opening a drawer.

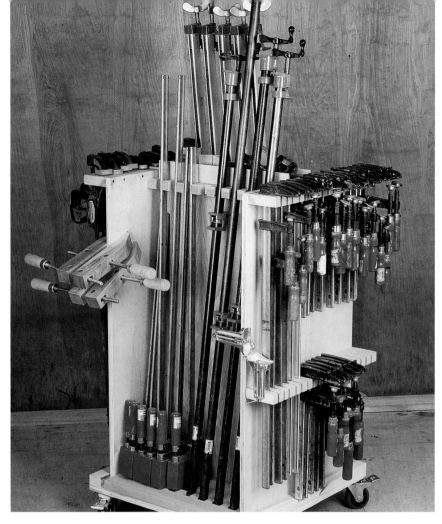

A mobile clamp rack saves wall space and puts the clamps where you need them: close by.

Lighting

No matter how organized your shop is, you cannot do good work without adequate light. How much light is adequate? A lot more than you'd probably imagine.

Lighting design standards for a cabinet-making shop suggest that it have 1,000 lux of illuminance—in practical terms, it should be at least as bright as a supermarket. For comparison, the suggested illuminance for general office space or kitchens is 500 lux. Moonlight measures about 1 lux.

How do you translate suggested lux into the number of fixtures you need in your shop? A lighting designer would use the lux number and work through several equations, taking into account the size and construction of the room, the wall and floor color, the fixtures' design, and other factors, and come up with a shopping list.

You can boil it down to this: Get one 4' double-tube fluorescent fixture for every 36 sq. ft. of shop space. If your shop walls are dark, or if the overhead is open-joist work with no ceiling, you'll need 50% more fixtures. When you do the math, round up and err on the side of more fixtures because aging eyes need even more light.

Position the light fixtures around your shop to avoid dark corners. Put a line of lights down each side of the shop to illuminate your benches and storage and hang the rest in the middle of the room. You may not end up with even spacing because of things like garage-door tracks, ductwork, or beams. Just work with what you have and position fixtures so the obstructions don't block too much light. Finally, use task lighting whenever you need a little more clarity.

Recycle empty drywall buckets or buy new ones, but keep a few on hand. Though you can get nifty organizers for carrying tools in a bucket, I prefer to use them for other things.

- Use them for storing small clamps—they're convenient for carrying them to the workbench.

- With the appropriate trays, they're a great way to store fasteners.

- Use them as a storage bath for waterstones.

- Manage extension cords by threading the pronged end out of a hole near the bottom and coiling the cord in the bucket.

- Store used solvents in them until your town has a hazardous-waste-disposal day.

- Filled with mineral spirits and kerosene, they can keep your best paintbrushes clean and ready to use.

- Filled with sand (or water) they serve as weights for clamping or veneering.

- Turned upside down, they make great seats or makeshift sawhorses.

The Efficient Shop

After spending time working in your Basic Shop, you're probably aware of its limitations. You've become an expert at ripping with a circular saw, but you dislike all the work needed to set up the sawhorses, extension cords, and foam for a few cuts. You look forward to stepping up to a tablesaw, locking the fence at the desired width, and ripping away without all that set-up time.

What to Consider

You're eager to get more time on woodworking by reducing the tedium and focusing on the fun parts. You've also built enough projects to begin to understand wood and its ways, and you probably want more control over your materials. You're ready to buy rough lumber and mill it to your own specifications so it's flat, straight, true, and dimensionally stable. Maybe you're even thinking about buying thick boards and resawing them into thinner pieces so all the wood in a tabletop or set of drawer fronts matches in shade, tone, and grain. If so, it's time to add some big machines to your shop.

Jointer and Planer Make a Pair

Good work starts with clear, straight-grained wood that's machined flat, and to do that you need both a jointer and a planer. Jointers flatten, straighten, and square up an edge, but they don't ensure that both faces are parallel to each other. That's why you need a planer. A planer simply renders the top of a board parallel to its bottom. In some cases it can remove twist or cup, but you can't count on it (see "Five Steps to Four-Square Lumber" on p. 64).

Tablesaw or Bandsaw Next?

After planing, a board is flat, uniformly thick, and has one good edge that's square to both faces. The next step is to run the good edge against a fence and rip the board to width on a tablesaw or bandsaw. Which should you have in your shop? Both do the job; the one you pick is more a matter of philosophy than fact.

Many novice woodworkers prefer the bandsaw because it's more forgiving: An error in guiding the work means a wavy saw kerf and more planing in the next step. A similar error on a tablesaw could result in a serious injury (see "Pros and Cons of the Bandsaw and Tablesaw" below).

I could make a strong argument in favor of either tool; in the end you'll choose the saw that suits your woodworking style. If you're partial to sculpted shapes, enjoy carving or turning, or

PROS AND CONS OF THE BANDSAW AND TABLESAW		
	Bandsaw	**Tablesaw**
Pros	■ No kickback ■ Cuts curves ■ Resaws ■ Does not require a fence or guide for every cut ■ Cuts angles safely and easily ■ Is quiet ■ Has small footprint	■ Rips cleanly (needs only a few strokes with a plane to smooth) ■ Makes accurate crosscuts ■ Cuts clean joints with the right jigs ■ Cuts dadoes and grooves
Cons	■ Ripping cuts need to be planed flat and smooth after sawing ■ Needs many (simple) adjustments ■ Small table sizes make handling sheet goods difficult	■ Is prone to kickback ■ Needs accurate adjustments for safety ■ Requires a fence or guide for every cut

plan to build a boat, you'll want to start with the curve-cutting bandsaw. If you're into the Arts and Crafts style, or want to build case goods, buy the tablesaw first.

Bottom line: Eventually you'll have both.

Add Other Tools as Needed

After you've acquired the machinery for dimensioning lumber, put away your cash (or plastic) for a while. Plan to buy the other tools, but not until you need them. For instance, you can put off the drill press for a while, but when perpendicularity becomes a crucial attribute for the holes you need to drill, it's time to get one. The same is true for the router table and the mechanic's tools.

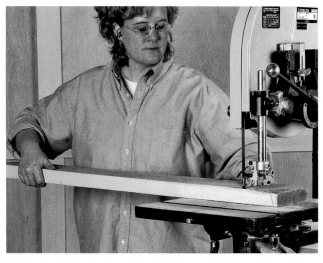

Resaw logs into lumber. The bandsaw can cut angles and curves, crosscut, rip, and resaw thick pieces of wood into thinner pieces.

A funny thing about getting a new tool—once you've learned to use it, you'll find it's indispensable. It becomes part of your problem-solving arsenal, and you can't imagine how you got along without it.

As your shop gets more efficient, it also gets more complex. Adding these tools means you'll need considerably more space, not only for the tools but also for the larger projects you'll undertake because of them. You'll have to think about where you'll use the tools and how you'll store them. Electricity is suddenly an issue, and you may have to upgrade your wiring or add new circuits. You'll need better lighting as well as enough heat and ventilation to keep you comfortable.

It will become an Efficient Shop, a place where work is streamlined to minimize the tedium and maximize the fun. But don't become so concerned with efficiency that you turn your woodworking into another source of pressure in your life. Woodworking shouldn't be about having the most powerful machines or how quickly you can complete your projects—those are the concerns of production engineers. Woodworking as a hobby is instead about slowing down, connecting with the wood, and challenging your hands, your head, and your heart.

ANOTHER WAY TO LOOK AT IT...

MACHINES USED TO MILL ROUGH LUMBER TO SQUARE
- Planer
- Tablesaw
- Bandsaw
- Jointer

MACHINES THAT CREATE JOINTS
- Tablesaw
- Drill press
- Router table
- Bandsaw
- Jointer

MACHINES THAT CREATE DECORATIVE EFFECTS
- Table-mounted router
- Tablesaw
- Drill press
- Bandsaw

FLATTEN FACE SIDE

Run one face over the jointer as many times as necessary until it's smooth. With proper jointer technique, that face will come out flat in length and flat in width with no warp, twist, or bow. Mark it as the face side.

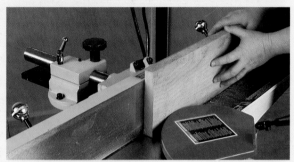

FLATTEN FACE EDGE

Place the face side against the fence (make sure it's 90°) and run one edge through the jointer until it cuts along the full length and full width of the edge. Mark it as face edge.

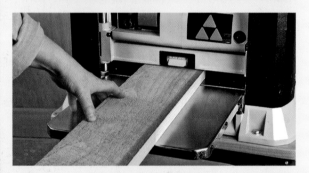

PLANE TO THICKNESS

Place the face side down on the table when passing the wood through the planer. It shaves the top until it's parallel to the face side. Make the last pass on the face side, because the planer leaves a smoother cut than the jointer.

RIP TO WIDTH

Put the face edge against the tablesaw or bandsaw fence and rip to the desired width. Take a few swipes with a handplane to remove the tool marks and prepare the edge for gluing.

CUT TO LENGTH

Use a miter saw or a miter gauge on the tablesaw to make smooth, square crosscuts to length.

Bandsaw

The bandsaw is the champion of cutting curves, but it can also rip and crosscut as well as cut angles, compound shapes, and joints. It's also the best way to resaw a thick piece of wood into two thinner pieces. In short, it's one of the most versatile tools you can own. It can do nearly everything a tablesaw can do, except cut grooves. And the bandsaw is a friendly, relatively safe tool that doesn't take up much room in a shop.

Bandsaw Basics

The "band" of the bandsaw is the flexible steel blade that wraps around two rubber-clad wheels. A motor turns the lower wheel, rotating so the blade moves downward at the point of cut, holding the workpiece on the table. The bandsaw will not pick up the workpiece and fling it across the shop.

Guides keep the blade from wandering, as does tensioning the blade. This is done by tightening a screw to raise the top wheel. This puts enormous stress on the saw's frame, and it must be strong enough to hold this tension. The traditional material for bandsaw frames is cast iron, a heavy material that also helps to dampen the vibration caused by all the rotating parts. These days the trend is toward lighter-weight frames that get their strength from welded web frames.

When you refer to the size of a bandsaw in inches, you're talking about three measurements. A 14" bandsaw has 14"-dia. wheels, its throat width allows just less than 14" for a cut between the frame and the blade, and the table is about 14" square.

WORK SMART

- Hook your shop vacuum up to the dust port during ripping and resawing.
- Watch the position of your fingers while cutting angles or tapers.
- Round stock can rotate, pinching your fingers.
- Cutting too tight a curve will cause the blade to twist in the guide, bind in the wood, and possibly break.
- A ticking noise usually means the blade will soon break.
- Overtensioning will wear out the saw's bearings prematurely.

Prize lumber from firewood. Most woodpiles contain some treasures—sweet-smelling fruitwood or beautiful burled or spalted pieces. Screw the log to a sturdy right angle jig and resaw it on your bandsaw.

What to Buy

Buy a 14" saw and you won't need to trade up unless you want to resaw boards wider than 12". Don't bother with smaller saws; they just don't have the capacity for ripping or resawing. A larger saw is always nice, but there's a big price jump up to a 16" or larger saw.

The standard 14" bandsaw can resaw boards up to about 6" wide. An optional riser block bolts between the upper and lower castings to increase the cutting height to almost 12". You'll want it sooner or later, so just get it when you buy the saw.

A 1-hp motor is standard on these saws; however, it can stall when cutting thick wood or resawing. Go for a 1½ hp from the start and save the trouble and expense of changing it later.

Every bandsaw has two sets of guides that support the blade both above and below the cut (one set is under the table). Look for guides that adjust easily, because you'll need to reset the guides whenever you change blades.

Blades for the Bandsaw

The right blade is crucial for bandsaw performance. The wrong blade will cut slowly, with lots of heat and dust, and it won't cut the curve you want. You'll end up with a variety of blades

Continued on p. 68

WHAT A BANDSAW CAN DO

■ CUT CURVES

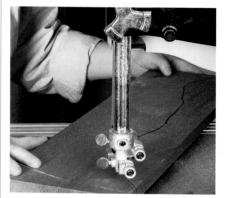

Cut open-ended curves and scroll.

■ RIP

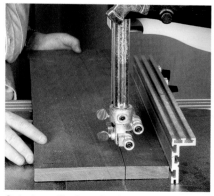

Cut along the length of a board.

■ RESAW

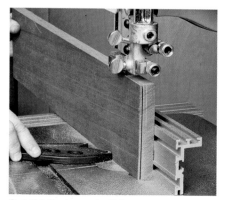

Cut thick pieces to thin.

■ CUT ANGLES

Cut angles and compound curves.

■ JOINERY

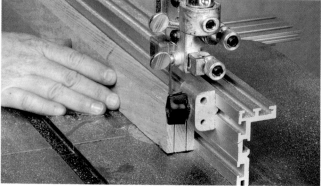

Cut joints like this tenon freehand or with jigs.

The 14" Bandsaw

The 14" cast-iron bandsaw is the workhorse machine in small shops worldwide, both amateur and pro.

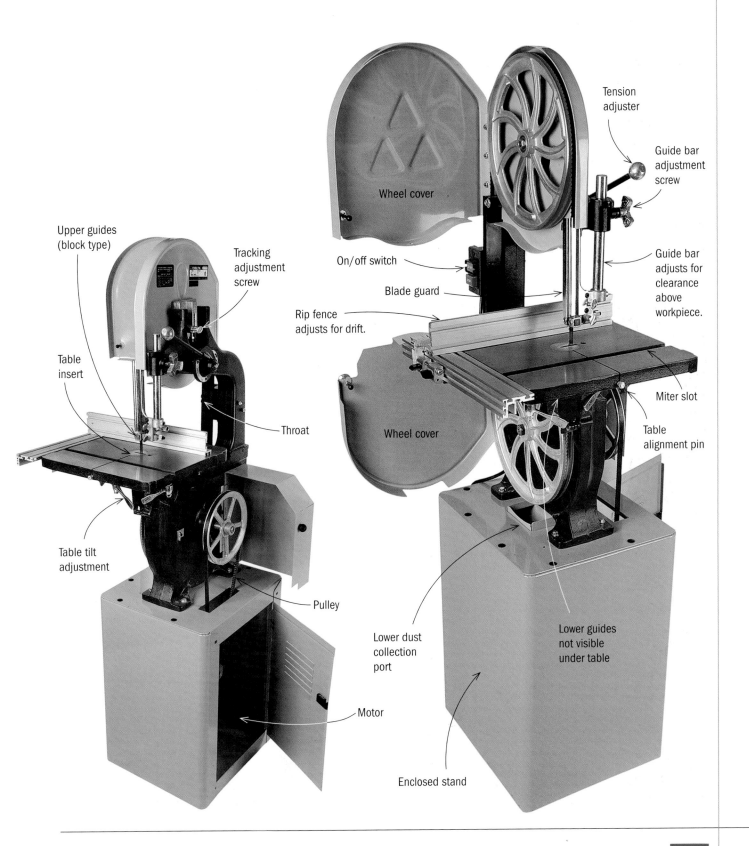

Tension adjuster

Guide bar adjustment screw

Wheel cover

Upper guides (block type)

Tracking adjustment screw

On/off switch

Guide bar adjusts for clearance above workpiece.

Blade guard

Rip fence adjusts for drift.

Table insert

Throat

Wheel cover

Miter slot

Table alignment pin

Table tilt adjustment

Pulley

Lower dust collection port

Lower guides not visible under table

Motor

Enclosed stand

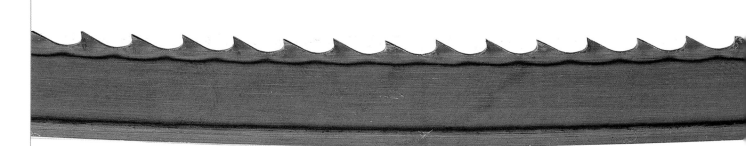

A typical blade. This ½" × 3-tpi bimetal hook-tooth blade is good for both resawing and sawing thick stock.

to suit the many jobs a bandsaw can do, changing them as the need arises. Here's what you need to know to find the right ones.

Select the right pitch

Pitch, the number of teeth per inch (tpi), is crucial to getting a smooth cut. The blade should have between 6 and 10 teeth in the work. For cutting a ½"-thick board, you'll use a blade with 12 to 24 tpi. If you're cutting a 2"-thick piece, use 3 to 5 tpi.

Tooth form

Bandsaw blades come in different tooth shapes. Regular tooth blades have a straight cutting edge and deep gullets. They're good for general duty. A skip-tooth blade also has a straight cutting edge, but there's a sharp angle between the tooth and the gullet. This makes for faster clearing of chips, so it's a good resaw blade. The hook-tooth blade has wider spaced teeth and the cutting edge is undercut. The front of the tooth swoops into the gullet in a sharp curve, making for an aggressive cut suitable for very hard woods.

Width

The wider the blade, the straighter it cuts. For the most part, stick with blades ½" wide or narrower. Wider blades require more force to tension than a small-shop bandsaw can generate. For cutting curves, you need a width suited to the desired radius. A ½" blade can cut a 2½"-radius curve; a ¼" blade cuts a ⅜" radius.

Material

Carbon steel is the most common and least expensive material for bandsaw blades. It's fine for general-purpose sawing, but it won't last long for resawing.

Bimetal and hardened blades use softer steel for the majority of the blade, with a band of harder steel at the teeth. They cost more and stay sharp longer.

THREE BLADES YOU NEED NOW

- ¼" × 20 carbon steel, regular, or skip tooth. For cutting thin wood and/or small curves.

- ⅜" × 6 bimetal or hardened steel, hook tooth. For general shop use in all but the thinnest boards.

- ½" × 3 bimetal or hardened steel, hook tooth. For resawing and general use in thick boards.

BALANCED WHEELS AND PULLEYS

Just like the wheels of a racing bike, the wheels and pulleys on your bandsaw should be well made and balanced for smooth rotation. Look for machined surfaces and feel the back of the wheels for spots that have been drilled out to balance the wheel.

BLADE TENSION

A sloppy blade will flutter and bow as it cuts. For best results, it must be adequately tensioned. As a rule of thumb for home-shop-size saws, tension your blades to the second-highest mark on the tensioning scale and don't use a blade wider than ½".

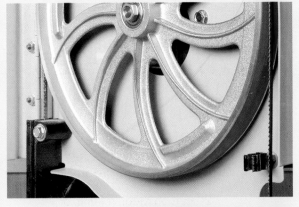

BLADE TRACKING

Once the upper wheel is tensioned, adjust its tilt to keep the band centered on the tire. Rotate the wheel by hand and turn the tracking adjustment screw (to the left of the spring tension scale in the second photo from the top) until the band settles in the middle of the tire. The tracking knob (to the left of the spring in the second photo from the top) tilts the upper wheel so the band runs around the middle of the wheels.

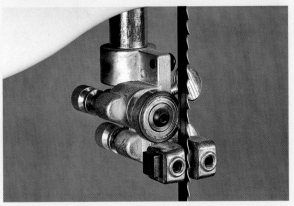

GUIDES

Limiting side-to-side and backward motion keeps the blade aligned and the cut precise. Adjust the side guides so they're close to the blade but not touching it. Guides take a lot of abuse and need frequent attention to do their job right.

To get the most benefit from the bandsaw's flexibility, you'll need to change blades often. It's not difficult with practice—the key is to do it in an orderly way. First things first: Unplug the saw.

PRELIMINARIES

First, release the tension, using the wheel or lever. Move the blade guides and thrust bearings out of the way. Then remove the throat plate and the level pin at the end of the table slot. Finally, ease the blade off the wheels and thread it through the table slot.

This is a good opportunity to do some basic maintenance. Clean dust and pitch from the tires. Also, clean the guides and make sure they're in good working order.

MOUNTING THE BLADE

While it may seem obvious, make sure you install the blade in the right direction. The teeth should face down toward the table. If they don't, the blade is inside out. Tension the blade and rotate the upper wheel by hand to test that the blade runs on the center of the tire. Adjust the tracking as necessary. Replace the throat plate.

ADJUSTMENTS

Adjust the upper and lower guides and thrust bearings. The thickness of a dollar bill sets the distance between the guide and the blade. The thrust bearing at the back of the blade should just kiss the back of the blade before cutting. It's wise to check the table for square rather than rely on the built-in protractor gauge.

If you use a rip fence, you'll need to adjust its angle to match the blade's cut. Choose a straight-edged piece of wood and draw a line parallel to the edge. Saw about halfway down the line (a little waviness is okay—the general trend is what matters) and stop the saw while keeping the board in place. Adjust the fence so it's against the board (see the photo below right). Finally, clean and wax the table.

Set the guides. A folded dollar bill is an inexpensive alternative to a feeler gauge for setting the correct distance between guides and the blade.

Reset the rip fence. Each time you change the blade, you'll have to adjust your rip fence to match the blade's cut.

Tablesaw

The tablesaw is the center of many wood-working shops. You may already know that it's the most efficient tool for ripping wood to width, but it does a lot more than that. Its ability to saw straight lines makes it a great way to cut grooves along a board's length, or dadoes across its width. If you need a big rabbet, a tablesaw can do it in two saw cuts. With a good crosscut sled running in the miter slot, the tablesaw is perhaps the most accurate way to crosscut or miter. When the face side of a board is run at an angle over the blade, it can cut an almost infinite variety of cove moldings. Add the ability to cut bevels, and you get a tool that not only cuts lumber to size, but also cuts complex joints and molds edges. No wonder it's so widely used.

But the tablesaw has a dark side—it does not tolerate errors. Ignoring proper technique or failing to attend to details can result in serious injury. Many beginning woodworkers fear the tablesaw with good reason, but acceptance is a better attitude. The turning blade's power is inexorable; you can't stop or change it, but you can harness it. If you understand the tool and use it on its own terms, you'll enjoy a lifetime of safe and successful sawing.

Stand to the left of the blade when ripping and use featherboards to hold the board against the fence. Feed the board steadily, pushing against the fence just behind the blade.

What to Buy

Get a 1½-hp contractor's saw with a mobile base, all the optional cast-iron tables, and the largest side extension table you can fit into your shop. You'll never regret buying capacity. A few brands offer optional cast-iron sliding tables, ideal for crosscutting; just keep in mind that they take up a lot more room.

For about half the money, you can get a 10" portable benchtop saw. They're at their best with surfaced lumber that's less than about 1" thick, and they will need babying to handle more. The small direct-drive motors are loud, prone to overheating, and easy to bog down. A benchtop isn't a bad saw to start out with—the price is right and it's not too intimidating. But you'll soon outgrow it.

■ RIP

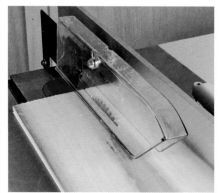

Run a jointed edge against the fence, and rip parallel to that edge.

■ CROSSCUT

Cut across the width, always guiding the cut with a device running in the table's slot.

■ CUT ANGLES

The blade angles to 45°. Position the fence so the blade angles away from it.

■ CUT MITERS

Using a miter gauge running in a slot, you can crosscut any angle.

■ CUT GROOVES AND RABBETS

Cut grooves, dadoes, or rabbets.

If you're flush with cash, consider getting a cabinet saw, the top-drawer choice. You'll pay twice as much as you would for a contractor's saw, but it will hold its resale value. With 3+-hp motors driven by two or three belts, these saws have the guts to saw through anything and do it all day without overheating. Their greater mass dampens vibration, and the tilt and depth mechanisms (bolted to the cabinet) are engineered to be tough, accurate, and easy to adjust.

On contractor's and benchtop saws, the depth and tilt mechanisms are bolted beneath the table. They're notably fussy, and it can be difficult to get and keep the blade parallel to the miter slot, especially after tilting the blade. Work around this by locking the blade in the vertical position and building sleds for cutting angles, as shown in the second photo from the top on p. 75.

No matter what saw you choose, get the fence upgrade, or buy an aftermarket fence. Include dust collection ports or skirts as appropriate, and get at least two blades—a rip blade and a 40-tooth combination blade.

Contractor's Tablesaw

A solid contractor's saw handles most of the tasks a woodworker will tackle, at about half the cost of a cabinet-style saw.

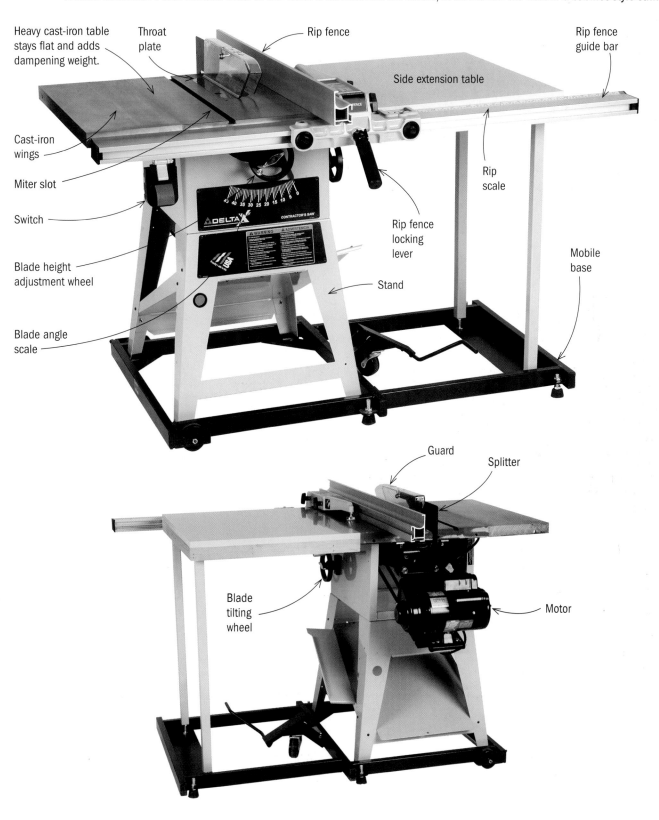

Heavy cast-iron table stays flat and adds dampening weight.

Throat plate

Rip fence

Rip fence guide bar

Side extension table

Cast-iron wings

Miter slot

Switch

Blade height adjustment wheel

Blade angle scale

Rip fence locking lever

Stand

Rip scale

Mobile base

Guard

Splitter

Blade tilting wheel

Motor

Benchtop tablesaw.
A 10" benchtop saw is capable of accurate but light-duty work. It's affordable, portable, and easy to stow under a bench or even on a shelf.

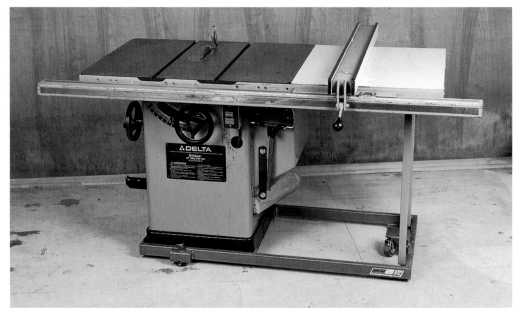

Cabinet-style tablesaw.
A 10" cabinet saw is heavier, more accurate, more powerful, and easier to adjust. With a mobile base, you can easily move it around your shop.

Safety

The tablesaw presents two distinct hazards: the exposed blade and kickback. The exposed-blade hazard is simple to avoid: Keep the blade covered and keep your hands away from it. Every new tablesaw comes with a blade guard; use it. If you find it clumsy or poorly designed, replace it with a better one. Keep your hands at least 6"

away from the blade and use push sticks and featherboards to hold and control the work safely.

Kickback is a more complex problem. It happens when the wood binds and catches on the blade's back edge. The rotating blade first lifts the wood up and, as the rotation continues, flings it back toward the operator. Sometimes the trajectory is low and the piece hits at hip

RIP FENCE

The tablesaw cuts parallel to the fence, so be sure to run a straight, jointed edge against it whenever ripping stock. A good fence should be so accurate you don't need to measure each cut but can simply read the measurement off the scale.

ANGLE SLED

Cutting bevels can be tricky, but a sled like this one makes it easy. By keeping the blade at 90°, the torsion bars in the undercarriage won't rack and cause misalignment. It also ensures accuracy and is safer.

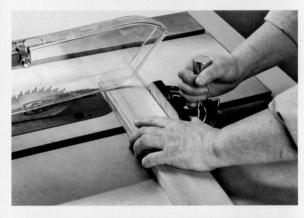

MITER GAUGE

The miter gauge slides in the slot and rotates to cut angles. For best results, attach a sliding auxiliary fence and adjust it close to but not touching the blade.

CROSSCUT SLED

The safest and most accurate way to crosscut is to place stock on a sled that rides on runners in the miter slots. A well-equipped shop accumulates several sleds—for small pieces, for 45° miters, for dadoes, for long pieces, for wide pieces, and so on.

Featherboards and push sticks guide the board and keep hands from getting too close to the blade. At the far right is a shopmade push stick. To its left is a plastic push stick. Aligned in the slot are a selection of featherboards: a wooden single-lock featherboard, a more secure double-lock variety, a shopmade featherboard that clamps to the table, and a magnetic featherboard.

level; sometimes it's at chest or head level. If you're standing in the right place (to the left of the blade), it'll pass you by, but others may not be so lucky. A kicked-back board moves at around 120 mph—more than enough to seriously injure someone on the other side of the room. Perhaps worst of all, if your hand is too close when the kickback starts, the initial upward motion can draw it into the blade. Kickback is scary, but you can prevent it by

using a splitter, setting up your saw properly, and using proper sawing techniques (see the photo on p. 71).

Ripping

- Use a splitter and guard.
- Run a jointed edge against the saw.
- Use a featherboard to hold the wood against the fence right at the blade.

- Every rip requires a fence, and only straight edges go against the fence.
- Every crosscut involves the miter slot in some way—don't use a fence or cut freehand.
- Always guard the blade; use a splitter when possible.
- Keep your hands away from the blade and use push sticks, hold-downs, and featherboards.
- Don't use the tablesaw when sickness, medications, anxiety, or fatigue might impair your mental agility.

- When sawing, push stock toward the back edge of the fence.
- Lean against the saw for stability.
- Don't reach; walk around the saw to collect your pieces.

Crosscutting

- Use a miter gauge, crosscut sled, or crosscut box.
- Never cut freehand.
- Never crosscut against the fence.
- Use clamps to hold your work against your sled or jig.
- Use a guard to reduce blade exposure.

Jointer

While it can bevel, taper, and even rabbet, the jointer's most important job is flattening the faces and edges of boards. A well-built, well-adjusted jointer removes twist, cup, bow, and crook during the crucial first two steps of properly milling lumber.

What to Buy

Jointers are precision tools and fussy to maintain. If the relationships between the tables, knives, and fence are not all perfect, the tool can't produce a flat surface. Your first concern when buying a jointer should be that the fit and finish are good enough to allow the necessary fine adjustments. Make sure the table and fence are flat—check them with a metal straightedge and reject a tool that can't pass this test. Smooth mating surfaces where the tables slide is crucial for fine adjustments, and the fence must move freely across the table's width.

WATCH OUT

- Keep your fingertips on top of the board—never hook them over an edge.
- Light cuts and slow feed leave a better surface.
- If the board won't cut full length or becomes tapered, the tables aren't adjusted correctly.
- Move the fence frequently to distribute wear across the length of the knives.

A long table handles lengthy boards with ease, and three knives cut more smoothly than two. Don't even consider a jointer with no outfeed table adjustments—it will be much harder to get and keep the proper adjustments so critical to proper operation. A cutterhead lock is a nice feature—it pins the knives at top dead center for easier adjustments.

WHAT A JOINTER CAN DO

■ FLATTEN THE FACE SIDE

Removes cup, twist, and bow from the face side.

■ FLATTEN/SQUARE AN EDGE

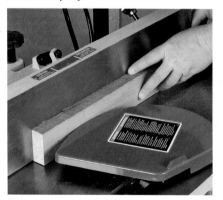

Removes bow and twist from the face edge and squares it to face side.

■ BEVEL AN EDGE

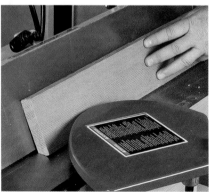

Angle the fence to bevel the edge of a board.

The 6" Jointer

Because it's the first step in milling four-square lumber, your jointer's size determines the maximum width board you can properly dimension.

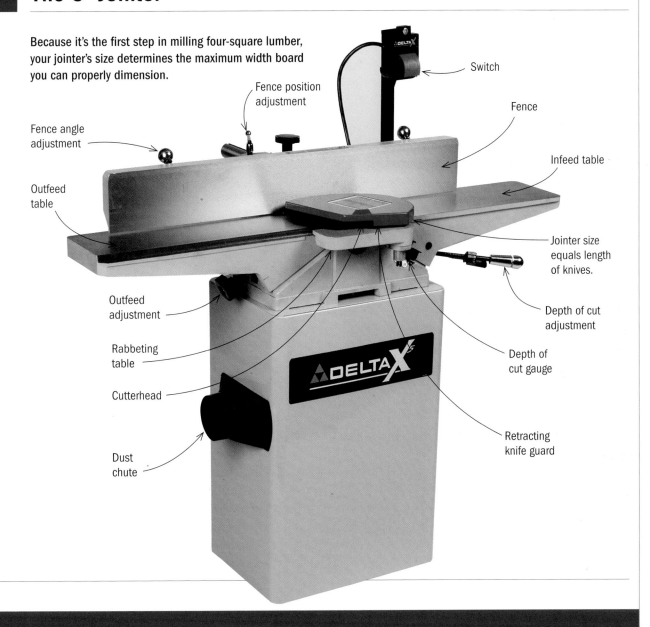

Switch

Fence

Fence position adjustment

Fence angle adjustment

Infeed table

Outfeed table

Jointer size equals length of knives.

Outfeed adjustment

Depth of cut adjustment

Rabbeting table

Depth of cut gauge

Cutterhead

Retracting knife guard

Dust chute

DELTA X

■ TAPER

A clamp defines the starting point to taper a long piece.

■ RABBET AN EDGE

Remove the knife guard to cut shallow rabbets.

Planer

A lot of people confuse the tasks done by a jointer with those done by a planer. A planer reduces a board to uniform thickness, with the top parallel to the bottom. It doesn't make the board straight by removing bow or twist—only a jointer can do that. To get a board both flat and square, you'll need both a jointer and a planer (see "Five Steps to Four-Square Lumber" on p. 64).

The planer is a simple machine and easy to use. As powered rollers pull a board over the bed of the planer, a spinning cutterhead removes material from the top of the workpiece. With proper knife and roller adjustment, you'll end up with a board that's of uniform thickness and smooth from end to end.

What to Buy

Get a portable planer that can handle boards 12" to 13" wide. For less vibration and smoother cuts, buy a machine with a cutterhead lock, and be sure to use it.

The number of times a knife cuts the wood also affects smoothness. The more cuts per inch (cpi), the smoother the surface. A low cpi means that the board moves quickly—a setting that's good for rough surfacing. For a finer finish, increase the cpi by slowing down the feed rate.

Sharp knives are the most important factor in smoothness. Buy a planer that makes changing blades as easy as possible, and change them often.

To keep your shop clean, get the optional dust collection hood.

WHAT A PLANER CAN DO

■ FLATTEN THE TOP

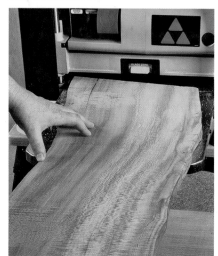

With the jointed face down, plane the top smooth.

■ REDUCE WIDTH

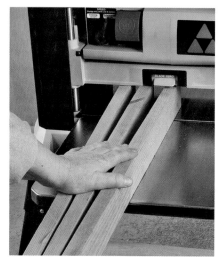

Reduce a number of boards to equal width.

■ REGULATE THICKNESS

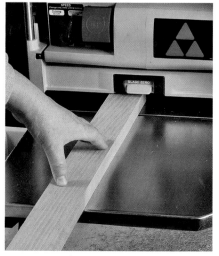

Plane a board to uniform thickness.

The Portable Planer

Once one side of a board has been jointed, the planer cuts a flat, smooth surface on the other side. It also ensures that the board has uniform thickness throughout its length.

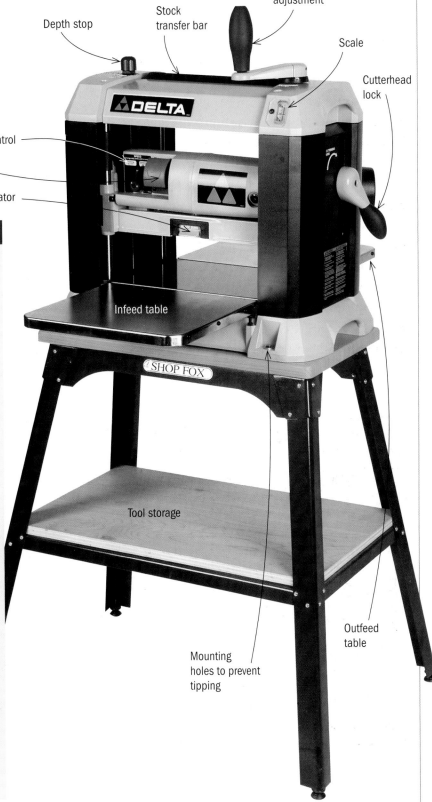

Depth stop

Stock transfer bar

Thickness adjustment

Scale

Cutterhead lock

Speed control

Switch

Zero indicator

Infeed table

SHOP FOX

Tool storage

Mounting holes to prevent tipping

Outfeed table

ALL TOGETHER NOW

Zero indicators and depth stops are useful features, but you can't count on having all the wood for a project the same thickness unless it's planed in the same session. Stack jointed boards on the infeed side and set the planer to take a light cut on the thickest board. Run every piece through, face side down, and stack them on the outfeed side. Carry the pile back to the infeed side, lower the cutter, and repeat as necessary. On the final cut, run the face side up to clean tool marks left by the jointer.

Planing it to size. The only way to ensure equal thickness is to plane all your stock at once.

Router Table

The router table is to the router as the tablesaw is to the circular saw. In both cases, mounting the tool upside down in a table and guiding the work over a protruding cutter produces something more versatile and functional than the original.

Most routing activities are easier to do on the router table (especially edge treatments), and you can rout pieces that are too small to do with the router in hand. A router mounted in the table can safely run big cutters with diameters too large to control by hand. A router mounted in a table also leaves smoother surfaces and cleaner profiles than one used on the workpiece. And if all that weren't enough, using the router table is faster because you're spared all that clamping and unclamping.

What to Buy

Your first router table should be a simple benchtop affair. It's small, easy to assemble,

---

WORK SMART

- Make sure the router is locked into the base before routing.
- Use a blade guard to reduce exposure hazard.
- Don't trap the workpiece between the cutter and the fence.
- Don't run a bit with a diameter greater than 1½".
- Make extra workpieces to test setups.

---

and can be stored under a bench when not in use. Just be sure to clamp it down to a stable surface when routing—otherwise, it might tip over. Look for one with a smooth, flat top that won't deflect under load and adjustments for leveling the insert plate with the top. Some router tables have blank inserts—you drill the mounting holes to fit your router. It's not a difficult process, but you save some trouble if you can get an insert that fits your router model.

WHAT A ROUTER TABLE CAN DO

■ MOLDINGS

Mold the edge of a board against the fence.

■ TEMPLATE CUTTING

Cut to a template using a bearing-guided router bit.

■ RAISED PANELS

A vertical bit creates the raised field of a raised panel.

The Benchtop Router Table

The router table is little more than a work surface that secures an inverted router on the underside of the tabletop. Add a fence and miter slots, and you've got a flexible tool that will get constant use in the shop.

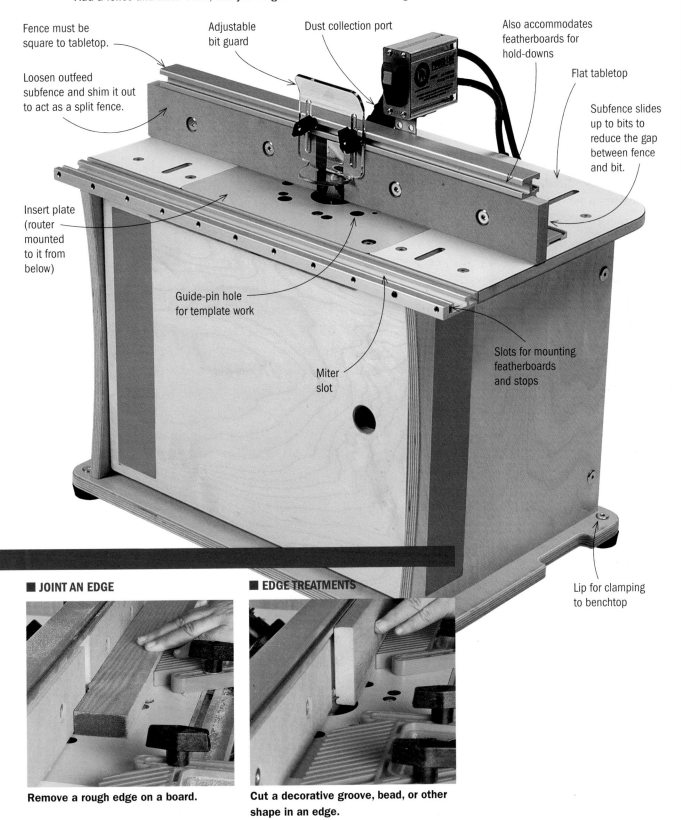

Fence must be square to tabletop.

Adjustable bit guard

Dust collection port

Also accommodates featherboards for hold-downs

Loosen outfeed subfence and shim it out to act as a split fence.

Flat tabletop

Subfence slides up to bits to reduce the gap between fence and bit.

Insert plate (router mounted to it from below)

Guide-pin hole for template work

Miter slot

Slots for mounting featherboards and stops

Lip for clamping to benchtop

■ JOINT AN EDGE

Remove a rough edge on a board.

■ EDGE TREATMENTS

Cut a decorative groove, bead, or other shape in an edge.

Drill Press

A drill press could claim a place in your shop if it merely drilled perpendicular holes. But it does so much more. It drills angled holes, runs a sanding drum or mortising attachment, and (unplugged) acts as a press for installing threaded inserts or tapping a hole. You'll use your drill press far more than you imagine.

Control your workpieces at all times by clamping them in place or restraining them against a fence. Otherwise, they'll rise with the bit and spin dangerously above the table. A fence is also useful for setting up indexes and stops for repetitive drilling.

What to Buy

Bench space is at a premium in most shops, so get a floor-mounted variable-speed drill press with a 16" to 17" throat. It has the power and versatility to handle just about any task you'll

WORK SMART

- Clamp workpieces down so they don't lift up when raising the bit.
- Never leave the key in the chuck.
- Always lock the table in place.
- Use a drill press vise for small pieces.

encounter. Get one with a ½" chuck to handle large bits.

Make sure the cast-iron table has a hole in the center for the drill to pass through and slots for clamping. In addition to whatever table is standard, you'll want to add a wooden table with a fence and good clamping arrangements.

A quill lock is a huge time-saver. With it, you can position the workpiece and pin it down by locking the bit in the full down position. This frees your hands to set stops, clamps, or indexing devices.

WHAT A DRILL PRESS CAN DO

■ DRILL PERPENDICULARLY

Get better results than you ever will by hand.

■ DRILL ANGLES

Tilt the table and lock it in place.

■ TURN A SANDING DRUM

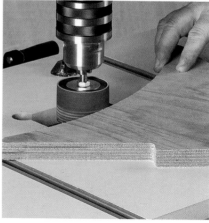

Sand curves and complex shapes.

16½" Floor-Mounted Drill Press

Whether you're drilling a quick perpendicular hole or clamping jigs to the table to make complicated joinery on furniture parts, a drill press gets steady use in the workshop.

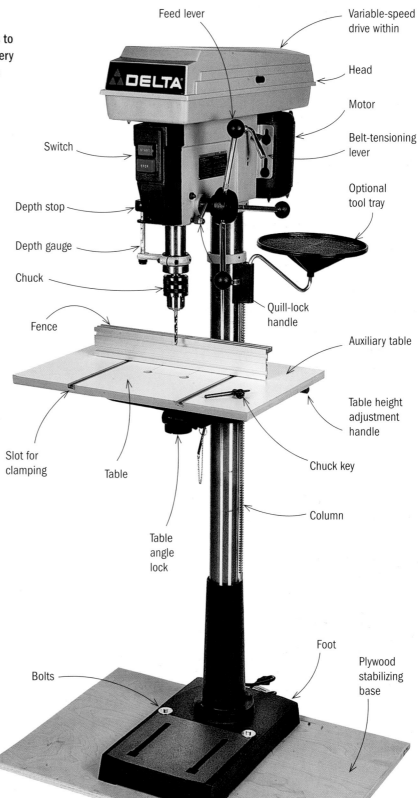

Feed lever

Variable-speed drive within

Head

Motor

Belt-tensioning lever

Switch

Depth stop

Depth gauge

Chuck

Optional tool tray

Fence

Quill-lock handle

Auxiliary table

Table height adjustment handle

Slot for clamping

Table

Chuck key

Column

Table angle lock

Foot

Bolts

Plywood stabilizing base

▪ MORTISE

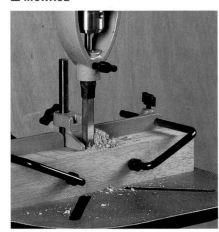

Use attachments to drill square holes.

Dust Collector

Good dust management requires a three-tiered approach. Your shop vac is the first line of defense. Use it for general cleanup and attach it to sanders and circular saws, as well as the miter saw, the bandsaw, and the router table.

Planers, tablesaws, sanding machines, and other big dust producers require a high-volume dust collector designed for woodworking machinery. The third line is a ceiling-mounted air cleaner to scrub out the tiny particles (down to 1 micron) that slip through the collector bags.

What to Buy

You can exhaustively analyze your dust-collection needs, but in a small shop there's no need to make it complicated. If you use a mobile collector and hook it up to only one tool at a time using no more than 10' of hose, get a 1½-hp single-stage collector (see the photo on the facing page). If your space, budget, and wiring allow, go for the 220v, 2-hp model. Most of the dust drops into the lower bag, while the permeable upper bag filters finer particles as the air blows through it. The standard upper bag only collects particles larger than 30 microns. Be sure to buy the optional 5-micron filter bag or one of the efficient new filtration canisters.

A typical home shop about the size of a two-car garage needs an air cleaner capable of moving 1,000 cubic feet of air per minute (cfm) (see the photo on the facing page). Make sure it has slower speeds as well—1,000 cfm is noisy. Opt for the washable electrostatic prefilter, and hold out for a remote control to avoid reaching for the switch constantly. Another great feature is a timer to turn off the cleaner after a couple of hours so you don't have to remember to do so.

WHAT DUST COLLECTORS CAN DO

■ COLLECT CHIPS AND DUST

Use large-diameter hose right up to the machine.

■ CONNECT

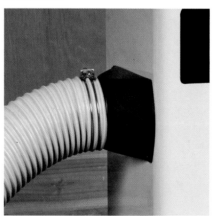

Use large ports to connect directly to machines.

■ SEPARATE

A trash can with a special lid keeps big chips and debris from damaging the fan blade.

Single-Stage Dust Collector and Air Cleaner

For most small shops, a 1½-hp single-stage dust collector is all you need. Move it from tool to tool as needed. An air cleaner will keep the shop noticeably cleaner. If your shop is connected to your house, it'll keep the house cleaner as well—less dust on the piano. If possible, hang it overhead a few feet from a side wall to promote air circulation.

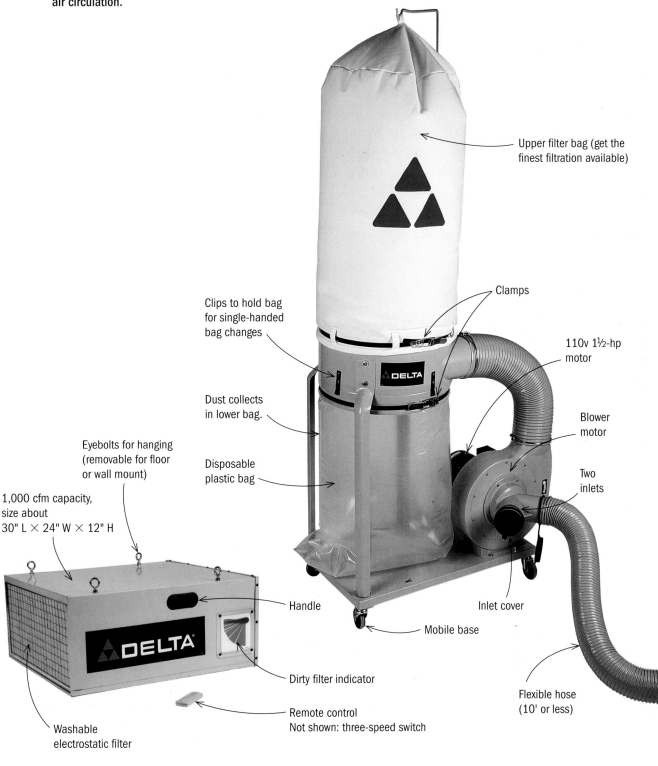

Upper filter bag (get the finest filtration available)

Clamps

110v 1½-hp motor

Clips to hold bag for single-handed bag changes

Blower motor

Two inlets

Dust collects in lower bag.

Eyebolts for hanging (removable for floor or wall mount)

Disposable plastic bag

1,000 cfm capacity, size about 30" L × 24" W × 12" H

Handle

Inlet cover

Mobile base

Dirty filter indicator

Flexible hose (10' or less)

Remote control
Not shown: three-speed switch

Washable electrostatic filter

Sanders

When you mention sanders, most people think of disk or random-orbit sanders, but there are a number of other sanding tools designed to tackle specialized jobs more efficiently. Admittedly, these sanders are not vital tools—it makes sense to buy them on an as-needed basis. But once you own any one of them, you'll find it solves all kinds of problems you didn't know you had.

Drum Thickness Sander

Like a planer, a thickness sander removes material from the top of the board. But rather than cutting with knives, sandpaper wound around the drum abrades the surface of the board as it moves on a conveyor belt.

A thickness sander with a cantilevered drum has a sanding capacity twice the length of its drum. You simply run one half of the workpiece under the drum, then flip it around and run the other.

The thickness sander aces some of the planer's biggest problems. First, there's no tearout—no matter how wildly figured the grain. Second, the sander is more accurate than a planer and is ideal for dimensioning stock to less than ¼". Third, the conveyor belt makes it safe to run small pieces through the machine.

What to Buy

Choose a sander with a cantilevered head for maximum capacity. Look for a large dust-collection port and casters or a mobile base. Securing the sandpaper to the drum is the most difficult part of using these machines; avoid any models that have fussy little clips tucked under the end of the drum. Remember that a cantilevered head is prone to sagging out of parallel to the table, and you'll have to check and adjust it often when sanding wide panels. The simpler the adjustment process, the better.

WHAT SANDERS CAN DO

■ THICKNESS WIDE BOARDS

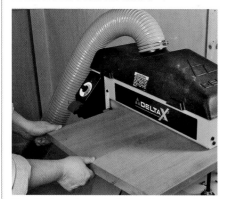

Handle boards up to twice the length of the drum.

■ SAND INSIDE CURVES

Smooth and fair inside curves.

■ SAND OUTSIDE CURVES

Smooth and fair outside curves.

The Efficient Shop Space

Adding all these machines to your shop complicates the space problem considerably. Once you move them in, the shop seems a lot smaller. And you can't just put your new machines anywhere. The tools have working relationships with one another, and they need to be located so that they can work together for maximum efficiency. If you're sharing your shop with cars, you'll have to figure out two locations for each tool—one for use and one for storage.

More machines means more load on your shop's electrical system, and you'll need circuits sufficient in number and size to handle the loads you'll put on them. Moisture is always a problem in home shops—too much or too little—and you'll have to monitor your shop and keep the levels in the proper range.

Finally, since you're spending more time in the shop, you need to think about how to keep it warm enough for your own comfort and safety.

Planning for Flow

A look at any coffee-table shop book proves that there's no one way to set up a shop. Organizing a shop is an ongoing process; as you gain experience and develop your own woodworking style, your shop will change until the space works best for you. For now, you can place your machines by thinking about the steps involved in processing rough lumber into workpieces.

The processing starts with long, heavy pieces of lumber coming off the storage rack for rough cutting to length. To minimize carrying, place the miter saw nearby (and remember that you'll have to crosscut really rough lumber with a circular saw or a handsaw). The lumber then goes through the jointer for jointing the face side and then the face edge. Locate the jointer so you don't have to negotiate corners or close spaces with a load of lumber and keep enough space around it to stack the wood during both operations. After the jointer, move to the planer.

■ CREATE ROUNDOVERS

Turn sharp edges into rounded corners.

■ SAND BEVELS

Tilt the table to sand a bevel with disk or belt.

■ REMOVE METAL

Use the belt or disk sander to modify metal tools and parts.

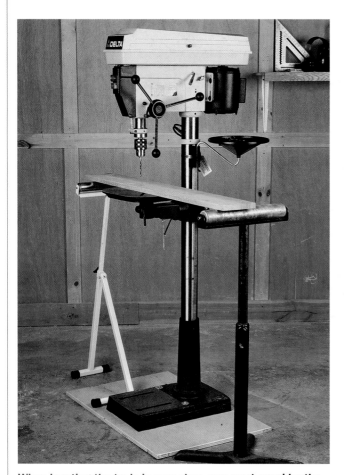

When locating the tools in your shop, **you must consider the infeed and outfeed tables. Make sure there's enough room around a tool to manage a long workpiece, here supported by roller stands.**

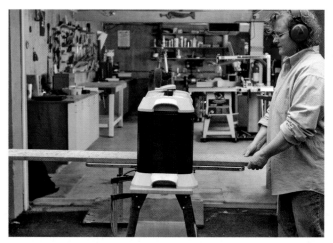

When working on **exceptionally long workpieces, you'll have plenty of room if you move outside.**

Few small shops have enough room **to allow the ideal infeed/ outfeed area around each machine. With mobile bases, you can easily move the machines to accommodate a large workpiece.**

From the planer outfeed, the stock goes to the tablesaw for ripping and then to the bench or to an active storage location—depending on the scope of the project. Active storage could be on the shelf under the bench, on sawhorses nearby, or back in the lumber rack.

To plan the flow in your own shop, start by measuring your shop and drawing a rough floor plan to scale. Include doors, windows, cabinets, posts, and other permanent structures. Next, fashion rough cutouts of your machines' footprints (to scale) and play around with positions and locations. Don't forget that the amount of space you'll need to use the tool is greater than its footprint—consider also the infeed and outfeed (see the photo above left). Think in terms of handling sheets of plywood and solid wood

up to 12' on a regular basis and larger workpieces on occasion.

Your floor plan won't show another important dimension—the heights of the tables—and this can be a major factor in machine placement. For instance, jointer tables are usually lower than tablesaw tables, so you can park yours in the tablesaw infeed/outfeed zone. You can even use it when sawing by fitting a shopmade bar of rollers to make up the difference in height.

In a small shop, a flow diagram might end up looking like a mad dance as the lumber moves through the milling process. That's okay—you'll move a few steps, work a while, and move again. The point is to minimize carrying, especially around corners or in confined areas.

Continued on p. 92

Floor Plan, the Efficient Shop

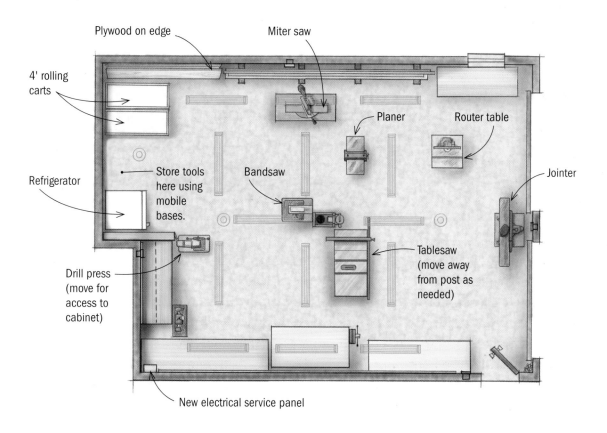

Plywood on edge

Miter saw

4' rolling carts

Refrigerator

Store tools here using mobile bases.

Planer

Router table

Bandsaw

Jointer

Drill press (move for access to cabinet)

Tablesaw (move away from post as needed)

New electrical service panel

For more options, store your machines on mobile bases. Settle on a layout for everyday use and shift the tools for unusual situations. You can even move machines outside when necessary.

The shop in the floor plan drawing on p. 91 is the usual setup in my shop. It clusters the machines around the immovable post. To free up space on the bench side, the tablesaw overlaps the post. On the rare occasions when I need to cut at full width, I simply move the saw nearer the bench. The bandsaw situation is similar—99% of the time it works there; if not, I move it. Located between the garage doors, the jointer has a little more than 9' of infeed and outfeed—plenty of room most of the time. The planer is the only machine that I regularly move—it stays against the wall until I need it. Though it's on a mobile base, the drill press rarely shifts location. On the base I keep a 5-gal.

bucket filled with sand to lower the center of gravity so that the top-heavy tool is easy to move when necessary.

I like to keep the machines around the edge of the shop and leave the middle open for assembly (it's also easier to get the cars in). But if I'm working on smaller projects and the cars stay outside, I sometimes move the jointer and the planer to the center of the bay and nestle the back of the jointer to the right side of the planer. In that location, they're ready for immediate use.

Upgrading the Circuits

Once you have an idea of where your tools will be used, you probably need to upgrade the electrical situation in the shop. After all, the space was not intended to be a shop. In most cases, you'll find that all the lights and outlets

FROM CONCRETE TO WOOD

Concrete is not the best material for a shop floor. It's hard on the feet and legs—you'll ache after a long day's work. It's rough on any tool you accidentally drop, and it tends to hold moisture so you can't leave wood resting on it for very long. A wooden floor takes care of all those problems, plus you can drive screws into it and use it as a laminating surface or for tacking down supports that stabilize projects during assembly. Here's a quick-and-easy method of installing a wooden floor that won't reduce headroom by much and is easy to remove down the road if you move.

Lay a sheet of heavy plastic over a clean concrete floor, taping any joints with duct tape. Put 2×4s down on their wide sides and fasten them to the floor with a powder-actuated nailer or Tapcon® concrete screws. Build a grid on 12" centers and fasten ¾" plywood or underlayment, with the joints falling on the centers of the 2×4s. Then leave it alone, or paint it a light color.

A sheet of ¾" plywood **makes an adequate ramp for the small difference in height between the shop floor and the driveway.**

If your shop is in the garage and you want to roll tools outside on occasion, don't run the plywood up to the edge of the 2×4 on the driveway side. Leave a ledge to support a plywood ramp (see the photo above).

CURRENT DRAW FOR COMMON WOODWORKING TOOLS

If the combined current draw of two tools exceeds the circuit breaker's capacity (typically 15 or 20 amps), you'll probably trip the breaker. Note the current draw figures are maximums—the tool will draw this much only under heavy load or when starting.

Tool	Typical Current Draw (Amps)
Circular saw	15
Random-orbit sander	3
Shop vacuum	10
Router	12
Miter saw	15
Jointer	8
Planer	15
Tablesaw	13
Bandsaw	9
Drill press	6
1½-hp dust collector	13
Mortiser	5
Drum sander	13
Belt/disk sander	9
6-gal. air compressor	18
Electric heater (220v)	20

Add a 100-amp subpanel in your shop and run your outlets on at least two circuits. To keep unauthorized users safe, you can turn off the circuits and lock the panel.

are on one 15-amp or 20-amp circuit shared with adjoining rooms. In that situation, running the bandsaw and the shop vac at the same time could trip the breaker and shut down not only the shop but also your kid's homework on the computer upstairs.

Lights, TVs, computers, and normal household items don't use much current and might never overload a circuit. Add a hard-working, power-hungry tool, and the combined current draw can easily exceed the capacity of the

wire, causing it to heat and possibly burn (see "Current Draw for Common Woodworking Tools" at left). The circuit breaker acts as a safety valve to spare the wire, shutting down when the current draw exceeds its limit. Simply installing a larger breaker is not a safe option; the wires are the weak link in the chain.

The best way to deal with this problem is to install a 100-amp line to the shop, with its own electrical panel and circuit breakers. While you're at it, add a 220v circuit for a heater and any larger tools you might add in the future. Run one 20A circuit down the bench side of the shop, another along the wall, with perhaps a third along the overhead and down the post. Install outlets every 10' or so and save yourself the clutter of extension cords. You can keep the existing lights and outlets on the household circuit, but label them as such and take care not to overload them.

With this setup, you can run more than one high-amp machine at once (planer and dust collector, for instance)—just plug them into separate circuits.

Building
Tool Storage

Everyone loves buying tools, and many even enjoy the sometimes finicky work of tuning them up. But with all those tools comes the need for tool storage that is safe and out of the way. A shop without a logical storage system is just a big pile of lumber and metal—not the kind of place where you can expect to get much work done.

In this chapter, you'll build projects that provide plenty of storage for the basic workshop, such as banks of drawers and cabinets with doors.

These projects are sized to fit a small shop, where space is at a premium. But the overall sizes of these projects are not critical. If you need larger storage cabinets, or a deeper case of drawers, adjust the sizes of the projects to fit your space and needs.

Many woodworkers feel that they should be building furniture or trimming the windows in their house instead of building storage units for their shop. While it's true that you can buy similar storage units at home centers, building your own storage units allows you to design the units to fit your needs exactly. It's also a good way to learn the skills that will improve your woodworking for building more critical pieces.

What You'll Learn

- Building no-fuss drawers
- Cutting dadoes with a router
- Installing a face frame
- Adding adjustable shelves
- Using butt hinges to hang doors
- Cutting mortises by hand
- Anchoring projects to wall studs
- Using a cleat system for hanging projects

The projects in this chapter are more than quick fixes for getting tools off the workbench. They're also designed to teach skills that are valuable to any woodworker. And wouldn't you rather learn how to build a chest of drawers using plywood than risk mistakes on expensive cherry or the irreplaceable walnut boards that came from a tree on your grandparents' farm?

The storage drawers in this chapter are a simple design. What's more, you don't need hardware; a drilled hole is a serviceable drawer pull and an interesting design element as well. You can use this same drawer design to outfit cases of all sizes and shapes.

Building the two-door cabinet is a crash course in cabinet and furniture making. When you learn to build a cabinet, you learn how to build a simple bookcase, add adjustable shelves, and apply a face frame. When it's time to hang the doors, you'll learn methods that work on everything from your home's front door to the doors in your bathroom vanity. You'll learn a clever method for hanging cabinets—or anything—on the wall.

A drawer for every need. Quick-to-make shop drawers offer lots of storage.

As you go, you may also want to build small projects necessary to keep any shop from being cluttered—racks for storing lumber and clamps and customized tool boards to help keep everything organized.

Throughout the process of building the projects in this chapter, you'll pick up ideas for building other storage systems and also learn a number of methods that will improve your woodworking skills. Not to mention, it's a good way to clean up your shop.

Case of Drawers

CASE CONSTRUCTION
The case of drawers shown here can be adapted to fit numerous uses around the shop or home.

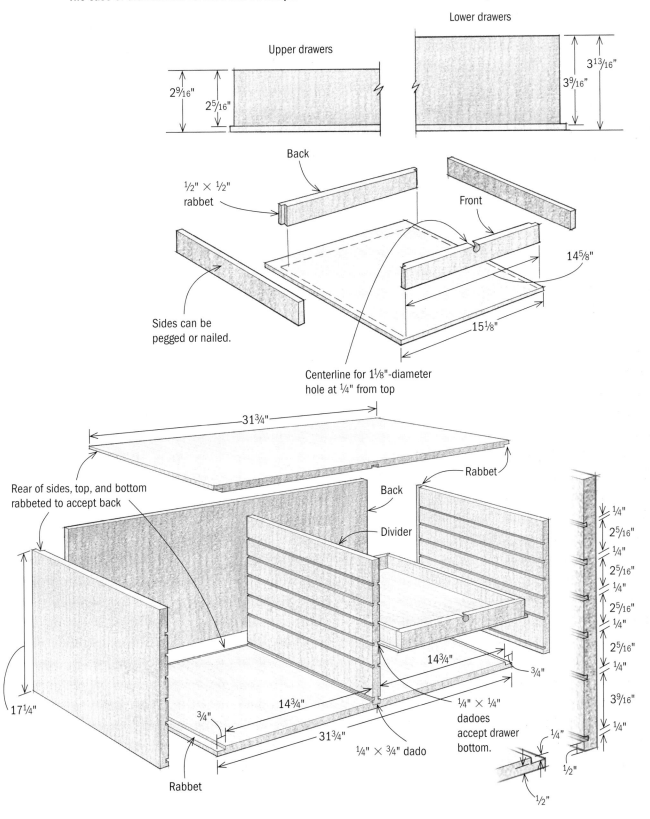

Upper drawers

Lower drawers

$2\frac{9}{16}$"

$2\frac{5}{16}$"

$3\frac{13}{16}$"

$3\frac{9}{16}$"

Back

Front

$\frac{1}{2}$" × $\frac{1}{2}$" rabbet

Sides can be pegged or nailed.

$14\frac{5}{8}$"

$15\frac{1}{8}$"

Centerline for $1\frac{1}{8}$"-diameter hole at $\frac{1}{4}$" from top

$31\frac{3}{4}$"

Rabbet

Rear of sides, top, and bottom rabbeted to accept back

Back

Divider

$\frac{1}{4}$"

$2\frac{5}{16}$"

$\frac{1}{4}$"

$2\frac{5}{16}$"

$\frac{1}{4}$"

$2\frac{5}{16}$"

$\frac{1}{4}$"

$2\frac{5}{16}$"

$\frac{1}{4}$"

$14\frac{3}{4}$"

$3\frac{9}{16}$"

$\frac{3}{4}$"

$17\frac{1}{4}$"

$14\frac{3}{4}$"

$\frac{3}{4}$"

$\frac{1}{4}$" × $\frac{1}{4}$" dadoes accept drawer bottom.

$\frac{1}{4}$"

$31\frac{3}{4}$"

$\frac{1}{4}$" × $\frac{3}{4}$" dado

$\frac{1}{2}$"

$\frac{1}{2}$"

Rabbet

	MATERIALS		
Quantity	**Part**	**Actual Size**	**What to Buy**
2	Case sides	$3/4" \times 19\,1/4" \times 17\,3/8"$	One sheet of $3/4"$ plywood is enough for all case parts.
1	Case top	$3/4" \times 19\,1/4" \times 31\,3/4"$	
1	Case bottom	$3/4" \times 19\,1/4" \times 31\,3/4"$	
1	Divider	$3/4" \times 18\,1/2" \times 17\,3/8"$	
1	Case back	$1/4" \times 17\,3/8" \times 30\,1/4"$	Two sheets of $1/4"$ plywood are enough for the back and the drawer bottoms.
10	Upper drawer fronts	$5/8" \times 2\,5/16" \times 14\,5/8"$	$1/2"$ or $3/4"$ also works for the drawer fronts and backs.
10	Upper drawer backs	$5/8" \times 2\,5/16" \times 14\,5/8"$	
20	Upper drawer sides	$1/2" \times 2\,5/16" \times 17\,1/2"$	
2	Lower drawer fronts	$5/8" \times 3\,9/16" \times 17\,1/2"$	You can substitute $1/2"$ plywood for the drawer fronts and backs.
2	Lower drawer backs	$5/8" \times 3\,9/16" \times 17\,1/2"$	
4	Lower drawer sides	$1/2" \times 3\,9/16" \times 17\,1/2"$	
12	Drawer bottoms	$1/4" \times 15\,1/8" \times 18\,1/2"$	
1 box	Finish nails	$1\,1/4"$	You'll only need a few, but it's best to buy a box so you'll have them on hand.
1 box	Brads	$1"$	
3	Drywall screws	$1\,1/2"$ long	
	Miscellaneous		Yellow glue, natural-colored putty (optional), wax, sandpaper, finish (if desired)

Buying Materials

One perk of using this method for drawer making is that no hardware is needed. For this project, I used three different thicknesses of plywood—$3/4"$, $1/2"$, and $1/4"$. But the plywood thickness is not important for the box back or the drawer sides and fronts, so you can use whatever scrap you have on hand. For the drawers, hardwood scraps work as well as plywood.

Adding a bank of drawers to the lower section of a workbench not only gives you much more storage space, it also adds weight to your bench, making it less likely to shift as you work. You'll find that the height of the drawers is perfect for shop storage, where most of your tools don't need a deep drawer. Chisels, screwdrivers, and some handplanes fit nicely into the shallow upper drawers, and drills and larger handplanes can be stored in the deeper lower drawers.

	TOOLS
■	Combination square
■	Tape measure or folding rule
■	Tablesaw with wooden auxiliary fence
■	Dado set
■	Drill
■	$1\,1/8"$ Forstner or brad-point bit
■	Hammer
■	Nail set
■	Circular saw and straightedge guide
■	Smaller block plane

But this drawer unit can be used in a number of different situations. Adjust the size to accommodate whatever space you have, but use the same straightforward methods to build the drawers. They're great in the shop, but at my house, we've got similar chests of drawers that we use in our laundry room and office.

Case of Drawers

Building the Box

Adjust the overall measurements of this drawer carcase to fit your workbench—or any other spot where you could use a case of drawers. You could use a number of different methods to assemble the basic case that houses the drawers, but here we'll assemble the box using rabbets and dadoes. On large casepieces, it makes the pieces easier to position and hold in place as they go together. You can also use biscuits or simply screw the case together.

Cut the parts to size

All the parts can be cut to size using your tablesaw, but if you don't have large outfeed tables, it may be easier to cut them to rough size first using your circular saw and a cutting guide (see "A Shopmade Cutting Guide" on p. 100).

1. When you measure for the height of the case, leave at least ¼" between the top of the drawer case and the bottom of the aprons on the workbench. You could aim for a tighter fit, but I find it safer to leave a little wiggle room in case either the case or the workbench is slightly out of square.

2. Most tablesaws aren't large enough to cut the top, bottom, and back to length, so use a circular saw and a cutting guide if necessary.

3. The sides, divider, and back can all be cut to the same height using your tablesaw.

4. Cut the top, bottom, sides, and divider to the same depth using your tablesaw. After making the cuts, adjust the fence and trim the depth of the divider another ¾" to allow room for the back.

Small space, lots of storage. This drawer unit provides a dozen drawers and plenty of storage for small hand tools and supplies.

Cut the joinery

All the joinery for this project is dadoes and rabbets. If you've got a dado set for your tablesaw, you can cut them easily, but you can also make the cuts using a handheld plunge router, a straightedge cutting guide, and straight bits. You'll need both ¼" and ¾" straight bits if you use a router. Here, we'll use the dado set on the tablesaw.

1. At your tablesaw, install the dado set and set it up to make a cut ¾" wide and ¼" deep. Install a wooden auxiliary fence on your tablesaw so that the blade won't nick the metal fence as you cut the rabbets. Then butt the auxiliary fence as close as you can to the blade without touching it.

2. To cut the rabbets on the ends of the top and bottom, butt the stock against the fence and pass the ends of the stock over the blade,

A SHOPMADE CUTTING GUIDE

A straightedge clamped in place allows you to cut a crisp line. It comes in handy for cutting large sheets of plywood to size. To make one, use ¼" plywood as the base and thicker plywood on the top as the guide. Just make sure the plywood used for the guide isn't so thick that it interferes with the saw casing when you cut. Once the base and guide are attached, saw through the base with the foot of your saw against the straightedge on the guide. Now, the edge of the guide's base will denote exactly the point where the blade cuts. The small guide shown here can be made in various lengths to accommodate different lengths of lumber.

To use the guide, line up the edge of the guide's base with the line you're cutting and clamp it in place. To cut, butt the base of the saw against the straightedge and keep it there throughout the cut.

CIRCULAR SAW GUIDE

A cutting guide is quickly made from scrap plywood around the shop. The base of the circular saw rides against the ¾" plywood, and the blade runs against the thinner ply below.

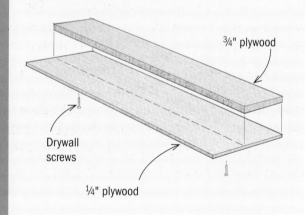

¾" plywood

Drywall screws

¼" plywood

Rabbet the top and bottom. **To rabbet the edges of the top and bottom, install a dado head set to ¾" and run the auxiliary tablesaw fence almost flush to the blade.**

Dado for the divider. **Shift the fence and set it to cut a dado down the center of both the top and the bottom.**

as shown in photo A. While you have this setup in place, cut rabbets on the rear inside faces of the top, bottom, and two sides to accommodate the back.

3. On the inside faces of the top and the bottom, mark a ¾"-wide dado down the center from the front to the back to accommodate the divider. Adjust the fence to make this cut and guide the top over the dado set, as shown in photo B. Use the same setup to cut the bottom's dado.

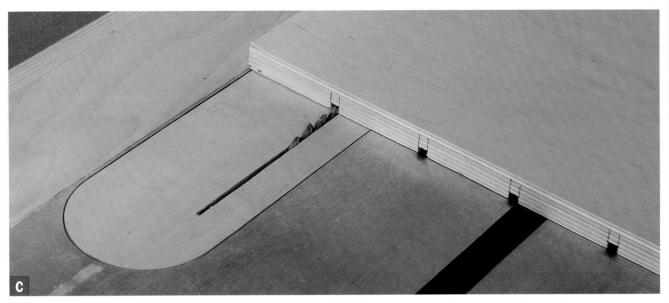

C

Dado using a single blade. **Instead of using a dado set to make the ¼"-wide dado, you can simply make two passes with a single blade.**

Cut grooves to accept the drawers

The drawers in this case are not traditional drawers, but they work just as well and are a little less fussy to make. Instead of installing drawer runners in the case, you extend the ends of the drawer bottoms to allow them to ride in grooves cut into the sides of the case. It takes only about 10 minutes at the tablesaw to cut all the grooves, or "runners," for the drawers. You can set up the dado set to cut a groove ¼" wide and ¼" deep, but I find it quicker to make every dado by taking two passes with a single blade.

1. When you lay out the dado locations on the sides and divider, don't forget that ¼" of the top and bottom will be buried in rabbets. It might help to put a mark ¼" down from the top to use as a reference point for measuring. Starting at the ¼" mark on one side piece, measure down another 2⁵⁄₁₆" and mark out a ¼" groove. At the bottom of the groove, measure another 2⁵⁄₁₆" down and mark out another ¼" groove. Do this all the way down the case (see the bottom drawing on p. 97). The height of the bottom drawer

is not critical, so if it winds up a little larger or smaller than the size shown in the drawing, don't worry. (The bottom drawer in this project is 3⁹⁄₁₆".)

2. Set the fence on your tablesaw to the marks for the first groove, then make the cut. Before you adjust the fence for the other cuts, cut the other side piece, as well as both faces of the divider. This way you won't have to measure the cuts on those pieces or reset the saw for those measurements later. If you're using a single blade, edge the fence over and make another pass to cut a groove that is ¼" wide, as shown in photo C. Using the single blade is often quicker than making all the test cuts with the dado set. Once you've made the cut, test the fit with a scrap of ¼" plywood to make sure that the plywood can move smoothly in the groove.

Skill Builder: Routing Dadoes and Rabbets

What You'll Need

- Plunge router
- Straight bit
- ¾"-thick scrap lumber at least 12" × 12"
- Scrap of plywood to serve as a straightedge guide
- Combination square

Cutting dadoes or rabbets is among the most common tasks performed by a plunge router. I know woodworkers who dedicate a certain router to that task alone, leaving it set up with a ¾" straight bit all the time. A router is a good choice for cutting dadoes and rabbets—it can be less hassle than setting up a dado set on the tablesaw for an exact cut.

To cut dadoes, rabbets, or grooves with a router, you need only a straight bit, sized to the desired width of the cut, and a straightedge guide. (I find a scrap of plywood to be the best choice for a guide.) But before you cut actual workpieces, it's good to practice the task on some scrap stock. The process described here is for cutting dadoes, but the same method works for cutting rabbets on the edge of a board or for cutting grooves.

1. Use a combination square to lay out the edge of the dado on the stock you're cutting, as shown in photo A. Make sure you mark both ends of the dado, so you'll have a reference for both ends of your straightedge guide.

2. With the router unplugged, install the straight bit into the router using the wrenches that came with your machine. Release the plunge lock, push down on the router, and lock the bit in place with ¼" of the bit exposed below the baseplate.

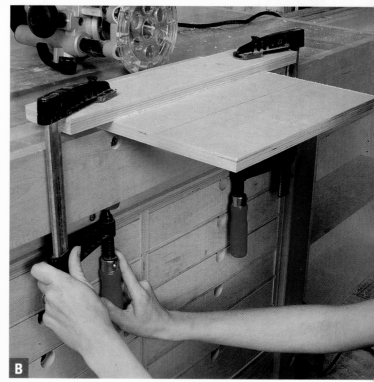

3. Clamp your workpiece to the bench. With the exposed bit hanging off the edge of the stock, align the bit with the marks for the dado, then butt the straightedge guide against the router to determine its location. Clamp the straightedge to the workpiece at that end. Then move to the other end of the workpiece, align the bit again, and set up and clamp that end of the straightedge, as shown in photo B.

Assemble the case

1. Before assembling the case, dry-fit all the parts to ensure that everything goes together smoothly and square. Then disassemble.

2. Set the bottom of the case flat on the workbench or another flat surface. Put a thin bead of glue in the dado and the rabbets, then set the sides and divider in place.

3. Put a bead of glue along the top of the divider and the sides, then set the top in place, as shown in photo D.

4. You can apply clamps to hold the assembly together until the glue dries, but assembly will go faster if you secure the top and bottom in place using 1¼" finish nails. Drive nails in place at the front and back of the top along the edges and on the divider, then use a straightedge to locate nails along the centers of the sides and dividers, as shown in photo E. Use a nail set to bury the nails about ⅛" below the surface. If you'd like, you can use a natural-colored putty to fill the nail holes.

4. Once both ends of the guide are clamped down, double-check the setup by measuring the distance between the dado marks and the edge of the guide at both ends. The two measurements should be the same. If they're not, you need to make an adjustment on one end or the other.

5. With the guide clamped securely in place, check the cutting depth of the straight bit and lock the plunge mechanism so that you can't plunge below the desired depth.

6. If your cutting depth is ¼" or less, you can make the cut in a single pass. Otherwise, you'll need to make several passes. Set the router on the workpiece against the guide, with the bit hanging over the left edge of the workpiece. Find a steady stance, get a good grip, then turn on the router and rout slowly from left to right, as shown in photo C.

Assemble the case. Add a bead of glue on the tops of the sides and the divider, and then add the top.

E

Why make things difficult? You can clamp the assembly together and wait for the glue to dry, or you can simply drive a few 1½" finish nails into place. For a clean look, use a nail set to bury the nails below the surface.

5. Once the top is nailed in place, flip the assembly over. Nail through the bottom into the sides and divider.

6. Flip the base over so that the front is face down. Set the back in place in the rabbets at the back of the case. Use a few finishing nails to secure it in place.

7. Stand the case upright. Measure the diagonals across the corners of the front to make sure the case is square. If the diagonal measurements aren't exactly the same, add a clamp across the long side and tighten until you get the correct measurements.

Building the Drawers

As far as drawers go, this construction method is among the most straightforward. The bottom of the drawer fits into the grooves on the sides and the divider. The drawer box is simply a rabbeted assembly that is nailed together. If you'd like a more refined look, you can use pegs instead of nails to assemble the drawers.

1. To make the drawer bottoms, measure and mark the width of the bottoms against the case. Leave about ¹⁄₁₆" on each side to allow the bottom to slide smoothly. If you leave more than that, the drawer might fall out of the grooves. On the tablesaw, cut the drawer bottoms to width.

2. Cut the drawer bottoms to the correct depth. Test the fit to make sure the bottoms line up flush to the front of the case when they're installed, as shown in photo F.

3. To make the drawer fronts, measure and mark the width of the fronts against the case. You want the fronts to be about ⅛" narrower than the opening on the case so that there's a little leeway on both sides of the drawer. Cut the drawer fronts to width on the tablesaw.

4. Measure and mark the depth of the sides against the case. You want the sides to be about 1" shy of the full depth of the case. On the tablesaw, cut the drawer sides to the correct depth.

5. Before cutting the fronts and backs to height, set up the tablesaw to cut ½" × ½" rabbets. Cut rabbets on the ends of all the drawer fronts and backs, as shown in photo G.

6. Hold the drawer fronts up to the front of the case and mark the height of each drawer. Remember that your lower drawers are a different height than the upper drawers, so you should measure and mark each one separately. Cut all the drawer fronts and backs to height, as shown in photo H.

Cut many rabbets at once. Rabbet along the edges of the fronts and backs to accept the drawer sides.

These drawers start at the bottom. Cut the drawer bottoms first, aiming for a flush fit at the front of the case.

7. After cutting the fronts and backs of the drawers, cut the drawer sides to the same height. You'll need to cut two sides for each drawer, making sure that the height matches the drawer fronts.

Cut the drawer fronts. Once the rabbets have been cut, cut all the drawer fronts to the correct height. Use this same setting to cut all the drawer sides. You'll need to adjust the settings for the lower drawers, which are a different height.

Two-Door Storage Cabinet

Shelving is great—it puts everything in sight and keeps it off the workbench—but a shelving unit with doors is even better. It allows you to store tools and supplies out of sight and keeps them from getting dusty as you work in the shop. While building this project, you'll learn how to add a face frame and doors, as well as how to outfit a cabinet for adjustable shelves. For more simple open shelving units, you can build the same unit without doors. The face frame can also be left off.

TOOLS

- Combination square
- Tape measure or folding rule
- Tablesaw
- Biscuit joiner
- Mallet
- Hammer
- Two clamps, at least 40" long
- Drill
- ¼" brad-point drill bit
- Drill press or drilling guide
- Handsaw
- Two sawhorses
- Dado set
- Marking knife
- Marking gauge
- 1" chisel
- Pencil
- A few dimes
- Level

MATERIALS

Quantity	Part	Actual Size	Notes
2	Sides	¾" × 8¼" × 32"	½ sheet of ¾" plywood
1	Top	¾" × 8¼" × 27½"	
1	Bottom	¾" × 8¼" × 27½"	
1	Back	¼" × 28" × 31"	
2	Face frame verticals	¾" × 1⅞" × 32"	Maple, poplar, or a similar hardwood
2	Face frame horizontals	¾" × 2½" × 25⅛"	Hardwood
2	Doors	¾" × 27⅛" × 12½"	
4	Butt hinges	1½" × 12½"	Brass or steel
8	#20 biscuits		Buy a containerful, you'll use them for other projects.
4	#0 biscuits		Buy a containerful, you'll use them for other projects.
1	Drilling guide	¾" × 3" × 26"	Plywood
2	Cleats	½" × 3" × 27"	Plywood
2	Shelves	¾" × 7½" × 27¼"	This design has two shelves installed, but you can add more if you wish.
8	Shelf pins	¼" dia. × ¾" long	Cut from ¼" dowel. You can also buy brass or plastic shelf pins.
	Finish nails	1½"	Buy a box so you'll have them on hand.
	Wood screws	#4	Use steel, not brass.
2	Friction catches		Screws for attaching come with the hardware.
1	Small block	¾" × 3" × 1½"	
2	Door pulls		
	Miscellaneous		Yellow glue

Two-Door Storage Cabinet

A face frame, simple plywood doors, and adjustable shelves make this cabinet a versatile storage spot for items you'd rather keep out of sight and away from dust.

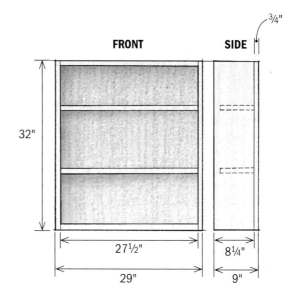

FRONT   **SIDE**

3/4"

32"

27½"

29"

8¼"

9"

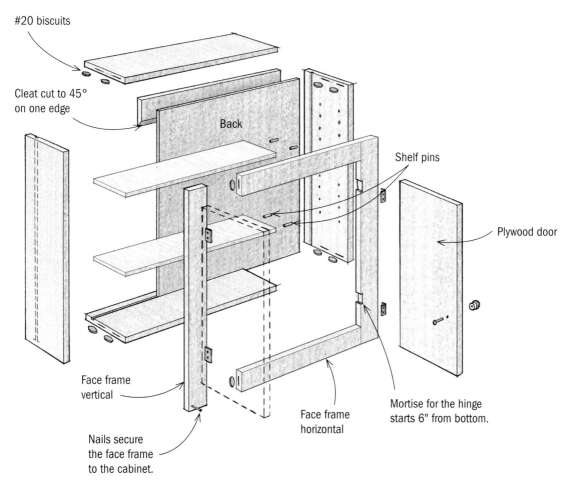

#20 biscuits

Cleat cut to 45° on one edge

Back

Shelf pins

Plywood door

Face frame vertical

Nails secure the face frame to the cabinet.

Face frame horizontal

Mortise for the hinge starts 6" from bottom.

Store it out of sight. **Whether you're making a small cabinet for storing finishing supplies or a larger version for power tools, the building methods are the same.**

Buying Materials

Chances are, if you've built another project you have enough scrap ¾" plywood on hand to build the basic cabinet and the doors. Maple was used for the face frame shown here, but poplar or pine would work as well. Just choose a wood that matches the color of the plywood. You can find the lumber as well as the hinges, friction catches, and wooden pulls at any home center. Inexpensive hardware works fine for shop projects.

Building the Basic Case

The basic case is built with ¾" plywood that is grooved to accept the back and biscuited together at the corners. You'll start by cutting all the stock to size and then cutting the joinery. Then, before you assemble the case, you'll drill the sides to accept shelf pins. This unit is designed to handle adjustable shelves, but if you'd prefer the shelves to be fixed, you can simply biscuit or nail them in place. If you're making open shelving, those are the only steps to make a shelving unit, but adding a face frame and doors will help keep out the dust. You can even install a lock on the doors.

Cut to size

1. Cut all the case pieces to size (see the drawing on p. 107) at the tablesaw or with a circular saw and cutting guide.

2. Before assembling anything, cut ¼" grooves in the top, bottom, and sides to accept the back of the cabinet, as shown in photo A. You can either use a dado set on the tablesaw or take two passes using a single blade. Inset the groove

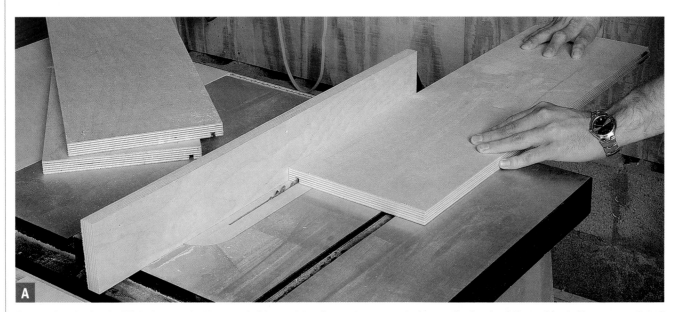

A

Groove for the back. **With the top, bottom, and sides cut to size, cut grooves to house the back of the cabinet. You can use either a dado set or take two passes with a single blade.**

½" from the back of the cabinet so that you'll be able to add the cleat later.

3. Trim the back of the cabinet to size. Remember that it needs to extend into the grooves at the back of the cabinet.

4. Cut the two cleats to size. Then cut the edge of each at a 45° angle. For more on this technique, see "Skill Builder: Hanging with a Cleat System" on pp. 118–119.

Cut the joinery for the case

1. Lay all the parts out on the bench and use a cabinetmaker's triangle (see "The Cabinetmaker's Triangle" at right) to mark the top, bottom, and sides.

2. Butt the ends of the sides against the bench stop or another 90° stop. Set your biscuit joiner

B

Biscuit the sides. Brace the biscuit joiner against a 90° stop and align the edge of the machine with the edge of the cabinet, then plunge to cut a slot for a #20 biscuit.

THE CABINETMAKER'S TRIANGLE

The hardest part of building a box—be it a bookcase, chest, drawer, or this storage unit—is often keeping all the parts straight. It's far too easy to confuse the side of a cabinet for the back or the side of a nearly square drawer with the drawer front. This can cause big problems when it comes time to glue up.

An easy solution is to mark project parts using a technique called a cabinetmaker's triangle. Put simply, it's a triangle broken into four parts. By marking each workpiece with a part of the triangle, you'll be able to tell at a glance where the piece you're holding belongs. Mark the front edges of your workpieces so you'll know which edge faces the front and so you can easily see all the marks when you're putting the project together. Bottoms of drawers or backs of cabinets can be marked with a full triangle. It's a handy and simple technique—all you need is a pencil—but it can save you lots of head-scratching and frustration.

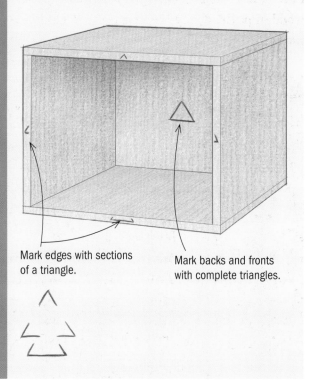

Mark edges with sections of a triangle.

Mark backs and fronts with complete triangles.

Cut slots in the ends. Lay the top and bottom flat on the bench, align the biscuit joiner with the edge of the piece, and plunge to make the cut.

to cut #20 biscuits. Mark out the biscuit locations or, since there are just a few slots to cut, save time and align the edge of the biscuit joiner with the edge of the case pieces and plunge-cut, as shown in photo B on p. 109.

3. Lay the top and bottom of the case flat on the benchtop. Align the edge of the biscuit joiner with the outer edges of the piece, then plunge-cut, as shown in photo C.

Drilling Holes and Making Shelf Pins

Making adjustable shelves may sound complicated, but it's not. In many cases, it's a lot easier than building fixed shelves. It's a good idea to have adjustable shelves in the shop to accommodate your tools and supplies as they grow and change. For adjustable shelves, all you need to do is drill holes in the sides of a case and insert shelf pins to support the shelves. To raise or lower a shelf, you simply move the shelf pins to another set of holes.

You can buy drilling guides, but making your own offers much more flexibility. I keep a number of drilling guides on hand around the shop—some have holes aligned at every inch, others at 2" intervals. I also make custom guides for specific projects.

You can buy brass or plastic shelf pins at hardware stores, but using lengths of dowels is both cheaper and better looking.

1. To make the drilling guide, cut a length of ¾" plywood to 3" × 26". Use a combination square to locate the centerpoint across the width of the board, then mark out holes at 1½" intervals. You can drill the ¼" holes using either a drill press, as shown in photo D, or a drilling guide set to 90°.

Make the guide. Use either a drilling guide or a drill press to drill 90° holes spaced 1½" apart and centered on the guide.

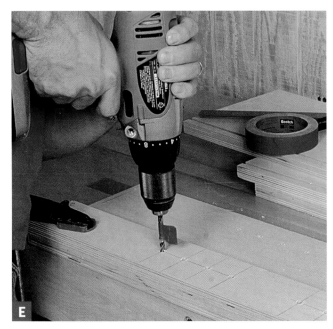

Drill for the shelves. Clamp the guide in place against the edge of the case side. Take care not to drill all the way through the side of the cabinet.

Insert the shelf pins. Once the cabinet is assembled, insert shelf pins into the holes to anchor the shelves.

2. To use the guide, align it with the bottom and side of a side case piece, then clamp it in place.

3. Before drilling, put tape on the drill bit at 1⅛" so you will not drill deeper than that. This amount allows for the thickness of the guide so you'll get a ⅜"-deep hole in the side. It will also prevent you from drilling through the side of the cabinet.

4. With the tape in place, drill the holes, as shown in photo E. Then clamp the guide to the other side of the case piece and repeat the process. Repeat again for the opposite case piece to get two rows of holes there. To secure a shelf, you'll need a total of four rows of holes.

5. To make the shelf pins, use a handsaw to cut pins to ¾" long. Test-fit the pins to make sure they fit well, as shown in photo F.

Assemble the case

1. Before gluing up the cabinet, dry-fit the parts without any glue to make sure you won't encounter any surprises during the final assembly. Then disassemble.

2. Lay one of the sides across two sawhorses or let it hang off either side of the workbench. Glue the biscuits into their slots and add glue to the biscuits and edges. Then set the top and bottom in place, as shown in photo G.

Set the ends in place. With one side of the cabinet laying across sawhorses, add glue and biscuits, then set the top and bottom in place.

Insert the back. Add just a few drops of glue into each of the grooves.

Secure the other side. Put the biscuits into place, then set the other side onto the assembly.

Clamp it up. A few clamps should pull the joints closed.

Check for square. Measure the diagonals across both corners. If the case is square, you'll get equal measurements for both diagonals. If not, adjust the clamps or add another across the long diagonal.

3. Put a small bead of glue in each of the grooves on the back of the cabinet and slide the back into place, as shown in photo H.

4. Glue biscuits into the ends of the upright top and bottom. Add a little glue to the biscuits and a small bead of glue on the plywood edges. Set the other side in place, as shown in photo I.

5. Apply clamps spanning from side to side to pull all the joints closed, as shown in photo J. You should see a small bead of glue bleed out of the joints. You can wipe it off now, but I find it easiest to remove excess glue with a chisel after it cures for about 30 minutes.

6. To ensure that the assembly has gone together square, use a tape to measure the diagonals from corner to corner, as shown in photo K. If the measurements are the same, the assembly is square; if not, adjust the clamps or add another one from corner to corner and tighten until the measurements are the same.

Adding a Face Frame and Doors

Whether you're building freestanding bookcases or built-in cabinetry, building and installing face frames is a task you'll encounter repeatedly. I've tried various methods for installing face frames—and there are almost as many methods as there are cabinetmakers—but the one outlined here is the easiest I've tried. I always use hardwoods for face frames, and I use #0 biscuits to ensure that all the front faces are flush to one another. Some people assemble the frames before they go on the case, but I find it easier to glue and nail the parts into place, one at a time, on the cabinet itself.

Make the face frame

1. Measure and mark a length of stock (at least 3½" wide) to the height of the cabinet. On the tablesaw, cut the stock to length, and then rip that piece into two pieces 1⅞" wide. These are the face frame verticals.

2. On the tablesaw, rip two other lengths of stock to 2½" wide. These will be the face frame horizontals.

Measure the frame members. With the two vertical pieces of the face frame cut to length and width, clamp them flush to one side, then hold the horizontal frame pieces in place and mark for length.

Mark for biscuits. With the frame members held in place, mark out centerlines for #0 biscuits.

3. Clamp the two vertical frame pieces flush to the edge of one side of the cabinet, then hold the horizontals in place and mark them to length, as shown in photo L. Crosscut the horizontals to length using a miter gauge on your tablesaw.

Skill Builder: Mortising for Hinges by Hand

What You'll Need

- 1" chisel
- Marking gauge
- Mallet
- Combination square
- Marking knife
- Pencil

There are numerous ways to cut mortises—with handheld routers, on router tables, or with specially designed mortising machines. But mortises for hinges are so shallow and small that using a router usually isn't worth the trouble because you have to fine-tune the cut with a chisel anyway. Learning to cut a mortise by hand teaches you to mark out for a hinge and shows you which measurements are critical. It's also a good way to improve your techniques with a chisel.

The method described here walks you through the steps in cutting a mortise for a butt hinge, but it comes in handy for mortising to fit other pieces of hardware around the shop.

B

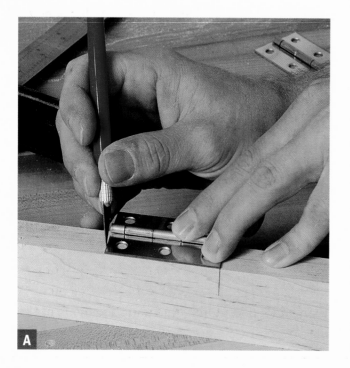

A

1. Set the hinge in place on the stock you'll be mortising. Hold the hinge steady and flush to the edge, then mark out the hinge using a marking knife, as shown in photo A.

2. Set a marking gauge just shy of half the width of the hinge barrel, then mark a line on both sides of the vertical frame member. The marking gauge may want to follow the grain of the wood, so take care to keep the guide pressed tightly against the side of the stock, as shown in photo B.

3. At the knife mark, place a chisel bevel-edge down. Moving down the width of the mortise, give the chisel light taps with a mallet to make a series of small cuts about 1/8" apart, as shown in photo C. Be sure not to cut below the mark on the edge of the stock.

C

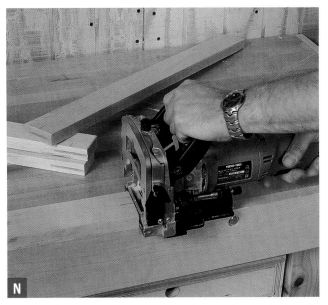

N

Biscuit the frame. With the frame members flat on the bench, set the biscuit joiner for #0 biscuit slots, then plunge to make the cut.

4. Set the face frame members flat on the benchtop or clamp them in place on the front of the cabinet, then mark centerlines for the biscuit slots, as shown in photo M (on p. 113).

5. Once the centerlines are marked, lay the frame parts flat on your bench and clamp them. Set the biscuit joiner for #0 biscuits and plunge to make the cut, as shown in photo N.

6. If you're adding doors to the case, it's easier to mortise for the hinges before you install the face frame. Measure and mark the mortise locations with a marking gauge and marking knife. It's easiest to cut shallow mortises like these by hand with a chisel. For more on this technique, see "Skill Builder: Mortising for Hinges by Hand" on the facing page. If you decide to add doors later, remove the cabinet from the wall and lay it on its side on the benchtop to cut the mortises.

7. Once the mortises are cut, you're ready to glue and nail the face frame in place. Start by nailing and gluing on one vertical frame piece, then add a biscuit and glue and attach one of the horizontal frame pieces, as shown in photo O.

D

4. With the bevel side of the chisel facing up, pare across the mortise until you reach the depth marked on the sides, as shown in photo D. You'll need to chop downward occasionally to reestablish the knifed lines marking the length of the hinge.

Nail on the frame. **Start applying the face frame by nailing on one of the vertical members. Add a biscuit and glue before attaching the horizontals.**

Apply glue along the edge of the cabinet, and add a few nails to hold the horizontal in place.

Make the doors

The doors on this cabinet are nothing more than two pieces of plywood, but when they're closed, you don't see any of the raw edges.

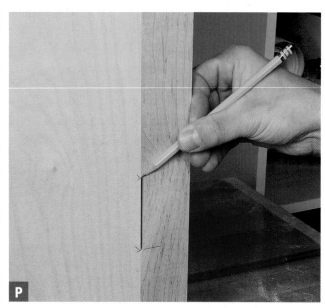

Mark mortises on the door. **Use a few dimes to shim the door into place, then transfer the mortise location from the face frame to the door. Cut the mortise just as you did on the frame.**

1. Measure and mark two pieces of ¾" plywood to fit inside the face frame. Aim for a gap on all sides of ¹⁄₁₆" (about the thickness of a dime). Cut the pieces on the tablesaw.

2. With the mortises cut in the face frame, you need to transfer the mortise locations to the doors, as shown in photo P. Set each door in place in the opening, using a few dimes as spacers to help you position it correctly. Mark the mortise locations directly off the mortises in the face frame and mark the depth using a marking gauge.

3. Use a 1" chisel to cut the mortises in the door, and follow the same methods you used for cutting the mortises in the face frames (see "Skill Builder: Mortising for Hinges by Hand" on pp. 114–115).

Hang the doors

1. Attach the hinges to the mortises in the doors, as shown in photo Q. Locate the center of the hinge barrel about ¹⁄₃₂" proud of the front of the door. Predrill the hole for the center screw, then install the screw. Make sure the inside edge of the hinge is parallel to the inside edge

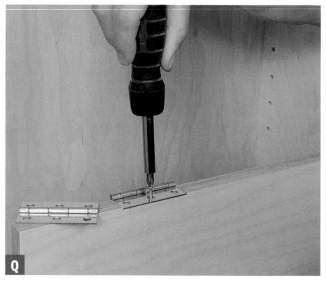

Attach the hinge to the door. **Before hanging the door, attach the hinge to the door with screws.**

R

Hang the door. **Align the hinges into the mortises on the frame, then screw a single screw in place. If the door doesn't close flush, remove the screw and start over. Don't add additional screws until you see that the door closes flush.**

of the door before predrilling the holes for and installing the other screws.

2. With the hinges installed on the doors, set them in place on the cabinet. Again, locate the center of the barrel about $\frac{1}{32}$" proud of the front of the door. Predrill and install only the center screw on each hinge.

3. With only one screw in each hinge, close the cabinet door and make sure it closes with the door flush to the face frame. If not, remove the screw in the hinge, make the necessary adjustments, and then install a screw in one of the other holes to test again. Once the door

closes flush to the frame, install all of the screws, as shown in photo R.

Adding the Finishing Touches

The cleat system used to secure this cabinet is as simple as it gets, but it's also the strongest way I know of to install a cabinet on a wall. After the cleat is installed on the back of the cabinet, you'll need only to add pulls to the doors and a friction catch to keep the doors closed. Once the cabinet is on the wall, you're ready to install the shelves and clean up shop.

1. Cut and install the cleats on the back of the cabinet and on the wall (see "Skill Builder: Hanging with a Cleat System" on pp. 118–119).

2. To add the friction catch, flip the cabinet upside down, then glue and clamp a small block flush with the face frame to the top of the inside of the cabinet.

3. Attach the female end of one catch to the small block about $\frac{1}{8}$" in from the top edge of the door, as shown in photo S.

S

Install a door catch. **Friction catches keep the doors closed. Glue a block to the frame. Screw one part of the catch on the block, then the other on the door.**

Skill Builder: Hanging with a Cleat System

What You'll Need

- Tablesaw
- $\frac{1}{2}$" plywood, 6" × 27"
- Screws
- Level

The cleat system shown here has been around for ages, and everyone seems to have a different name for it. I've often heard it called a French cleat, but I once heard a woodworker refer to it as an Indian cleat. "It depends," he told me, "on how you viewed that war."

Either way, it's a simple method that works beautifully. You cut one piece of stock in half at a 45° angle, then mount one piece on the back of the cabinet and anchor the other to the wall. To hang the cabinet, you just lift the cleat of the cabinet over the cleat on the wall, then lower it on to the other. The two cleats lock together, holding the cabinet snug to the wall.

1. Cut a length of stock 6" wide and as long as the piece you're hanging. Then set your tablesaw to cut at 45° and cut down the center to create two cleats, as shown in photo A.

A Cleat System

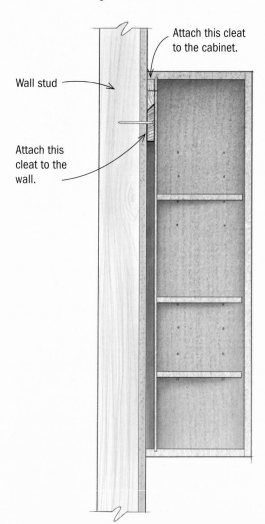

Attach this cleat to the cabinet.

Wall stud

Attach this cleat to the wall.

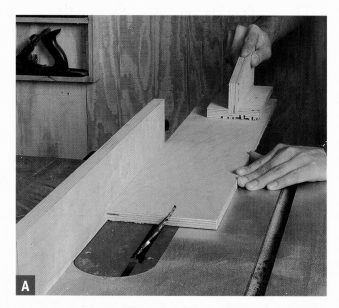

A

B

Install the shelves. **With the shelf pins in place, the shelves simply rest upon them.**

2. Attach one cleat to the cabinet using glue and drywall screws. Make sure the angled edge is facing down and that the long, full side of the stock is facing the wall, as shown in photo B.

3. Attach the other cleat to the wall, with the shorter angled side against the wall, pointing up. Attach one screw, adjust the cleat so it's level, and then add the other screws, as shown in photo C. Make sure that at least one screw goes into a wall stud. On larger cabinets or cabinets holding a lot of weight, you'll need more than one screw in a wall stud.

4. To hang the cabinet, lift it to the wall, raise the cleat of the cabinet over the cleat on the wall, then lower the cabinet into place. The two cleats lock together and secure the cabinet in place.

4. With one door closed and the other open, position the catch on the inside of the closed door so that it is in line with the female end of the catch on the small block when the door is closed flush, and mark the location. If you center the screws in the holes on the catch, they leave a little room for adjustment after they're installed. Repeat steps 3 and 4 for the other door's friction catch.

5. To install the door pulls, predrill holes about 8" up from the bottom of the door and about 2" in from the edge. Then set the screws in place from inside the doors and thread the pulls onto them.

6. Hang the cabinet on the wall. Add the shelf pins and put in the shelves, as shown in photo T. Now you're ready to load up the cabinet with tools and supplies.

Materials for Projects

When you're eager to start a project, taking the time to get the right materials can seem like a drag. Eager to get going, you might be tempted to use whatever is at hand. Don't give this part of the building process the short shrift—the wrong materials can cause big problems. If you choose wrong, you'll have trouble fitting the joints or perhaps will spend way more time than is necessary finishing (or refinishing). Worst of all, using the wrong materials can weaken your project and lessen its durability—it may come apart long before its time.

So, take the trouble to learn about your materials and look forward to successful woodworking with a minimum of fuss and bother.

Solid-Wood Lumber

Picking the right wood for your projects can make the difference between success and failure. If you buy boards that are bowed, warped, or otherwise misshapen, you'll have a hard time getting good-fitting joints. And even if you do manage to make things go together, your work will lack the visual crispness that comes from straight lines.

Good results start with good lumber. Here's what you need to know to find it.

Rarely does a tree grow perfectly straight and upright. It could grow in all kinds of crazy situations—on a hillside, around fallen neighbors, or curved by the wind or from seeking the sun. Every cell of the tree bears the imprint of the stresses that shaped it. Yet when it comes time to make boards, every log is treated the same and run straight through a saw, sliced into uniform slabs, and stacked in a kiln to dry.

In the kiln, a complex program of heat, steam, and pressure evaporates the sap and moisture from the boards and reduces the moisture level to about 10%, which is the average moisture level that well-seasoned wood maintains over the years. As the moisture content falls, the cells shrink, but not uniformly, and the boards twist, cup, crook, and bow to varying degrees, depending on the species, the thickness, and how and where the tree grew.

> ## WORK SMART
>
> **T**he wider the board, the greater the dimensional change with changes in humidity. As a rough rule of thumb, figure that a 16"-wide board will shrink or swell about 3/16" from the most humid season to the driest.

Defects in Solid Wood

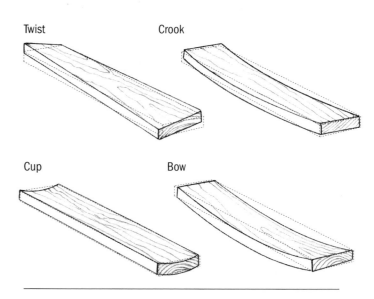

Twist   Crook

Cup   Bow

Once out of the kiln, the moisture content of the wood maintains equilibrium with the surrounding atmosphere. If you live in Winslow, AZ (one of the driest places in the United States), the wood will continue to lose moisture until it reaches about 7% moisture content. If you're in Lake Charles, LA (where the relative humidity averages 78%), the moisture content will rise to about 14%. Wherever the wood is, it will gain or lose moisture depending on the humidity of the air around it.

Each board gets bigger when the humidity is high and shrinks when the humidity goes down. That's why drawers are more difficult to open in the summer but glide smoothly in the dry winter months. Just how a piece of wood changes shape depends on where it was cut from the log (see "How a Board Acts Depends on Where It Comes From" on p. 126).

Reading the Grain

Upon leaving the kiln, the boards are milled flat and made into the rectangular planks you see at a lumberyard or home center. Twist, cup, and bow are removed in the process, but not always completely or for good. As the boards shrink and swell throughout their lives, they will continue to change shape according to their place in the original log.

When buying lumber, your first challenge is to find boards that are flat, straight, and square. Then you must figure out which boards will retain those characteristics once you get them to your shop. The clues are in the grain. Learn to read the grain, and you maximize your chances for woodworking success.

End grain

Check first for vertical end grain to find the most stable quartersawn boards. If the boards are narrow, select those with no pith. You might be able to find a wider board and rip away the pith using a tablesaw.

If there aren't enough quartersawn boards to complete your project, look for riftsawn stock, where the grain runs at around 45 degrees. These boards don't shrink much in width or thickness, but they do tend to become slightly diamond shaped.

Finally, select the best flatsawn boards, those with the flattest growth rings from the outsides of the trees. They cup less than boards cut nearer the center of the tree.

Face grain

Flatsawn boards have the familiar cathedral pattern along their length. Look for boards with a regular pattern, which indicates that the boards were cut from a fairly straight tree.

Reading End Grain

Riftsawn—the grain is approximately 45° to face.

Ideal—quartersawn

Flatsawn board from the outside of a large-diameter tree

Avoid! Cut too near the pith, this board will cup severely.

Reading Face Grain

Regular cathedral pattern

Multiple ellipsoids

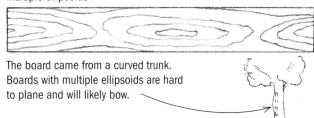

The board came from a curved trunk. Boards with multiple ellipsoids are hard to plane and will likely bow.

Quartersawn boards don't show cathedral grain on their faces.

Crook at runout (exaggerated for clarity)

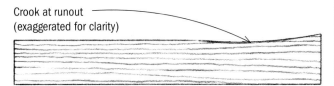

Ray fleck **shows in the quartersawn boards of some species. This is quartersawn white oak.**

The sapwood near the edge **of this cherry board is not only different in color but also less stable.**

Avoid flatsawn boards with multiple ellipsoids in the cathedral pattern. These occur when a straight board is sawn from a crooked tree. The internal stresses in the board will cause bow, twist, and other problems. It's a useful fiction to imagine that a board contains a "memory" of the whole tree and will always try to assume the shape the tree had in the forest.

Quartersawn boards don't have cathedral figure. Their faces show bold straight lines, sometimes running off the edge. That shows the tree grew with some curve; the board will not remain straight where the grain runs out.

Prized in large part for their dimensional stability, quartersawn boards of some species are spectacularly beautiful. Because of the cellular structure of the wood, quartersawn boards show ray fleck, the amount and style depending on the species (see the photo at top left). Quartersawn oak surfaces have shimmering rays interspersed with the strong grain, and many other woods, such as maple, show a pleasing mottled appearance.

Look for abrupt color changes at the edges of a board that indicate the presence of sapwood (see the photo at top right). Some people reject sapwood because they don't like the look of it, but the real problem is that sapwood is

Sight down a board **to check it for crook, bow, and twist.**

more sensitive to changes in relative humidity. It's less stable than heartwood.

Knots occur where branches grew on the tree and are not necessarily grounds to reject a board. Sometimes a knot can be an interesting design element, but knots can cause localized problems. Knots are harder than the surround-

WORK SMART

Watch out for black knots. Unlike the brown ones, black knots fall out. Sometimes when a tablesaw or router hits a black knot, it shoots off like a bullet.

ing wood, so they can make it difficult to flatten a surface. Because they shrink differently from the surrounding wood, knots can crack and split, and if they're on the edge of a board, they can cause a severe localized kink.

Many boards have splits at the ends, termed checks. They're not a problem unless they extend far enough into the board to effect yield. Just cut them off.

Edges

When you've found a piece of wood that passes both the end-grain and face-grain inspections, turn it on edge. You're looking for three things here—crook along the edges, bow along the face, and twist.

A small amount of crook is normal and easy to remedy, but a severe case should be rejected. A severely bowed board should be ditched, but a small amount of bow is not a problem, especially if the board is crosscut into smaller pieces. Over a shorter length, the bow becomes negligible.

Twist is difficult to assess accurately in the yard, but you should be able to learn something by looking closely at the far end. If the bottom of the board is off to one side as you sight down it, you've got some twist. Like bow, it's not a problem if the twist is not severe and the board will be crosscut.

Where to Buy Solid-Wood Lumber

The projects in this book use standard sizes of lumber readily available at virtually any lumberyard or home center. You'll find a good selection of lumber that's been carefully graded, selected, and milled into standard dimensions so it's easy to ship, store, and sell. It's not difficult to find decent lumber. The hard part is knowing how it's sold, what to ask for, and how to talk to the sales staff.

Most of the wood sold at a lumberyard or home center is rough structural lumber for building houses and decks. Some of it is wet—not even kiln dried, and what has been kiln dried is often dried to only 19%. There's no reason to dry it any further, since many yards store the lumber in piles under the open sky. It's not worth the expense to dry the wood when in the course of building the house or deck, the lumber will be exposed to the elements.

In the sheds you'll find the dimensional lumber, hidden from the elements and kept warm and dry. This is where you'll find the woodworkers. Here you'll find neatly stacked piles of pine, poplar, oak (usually red oak), fir, maple, cedar, mahogany, and sometimes cherry. The selection varies from region to region, but these are nearly universal.

Dimensional lumber is surfaced on four sides and is available in standard widths of 2", 3", 4", 6", 8", 10", and sometimes 12". Generally, these are nominal dimensions, meaning that the actual dimensions are less. A standard 1×6 starts out as a roughsawn 1×6. By the

WORK SMART

Always choose a little more lumber than you need to account for defects in lumber and mistakes. Most pros buy 15% extra; as a beginner, you might want to get a little more.

Boards flatsawn from the outer edges of a log shrink moderately in width and cup slightly, with the bowl away from the center of the tree. It's a useful fiction to think of the growth rings as trying to straighten out. Boards flatsawn from near the center of the tree shrink a little less in width but cup more; you could say the smaller-diameter growth rings have a stronger tendency to straighten out.

When a board is cut through the center of the tree, the growth rings appear as vertical lines at the outer edges and curved lines near the pith, or center, of the tree. If you saw out the pith from the center, you end up with two boards with vertical growth rings, which are said to be quartersawn. These boards shrink a little in width, a little less in thickness, and won't cup. Every pile of flatsawn boards will contain a few that are quartersawn,

but some sawyers cut to maximize the yield of quartersawn boards by sawing around the log. This requires more labor than flatsawing, but because quartersawn boards are so stable, they can be sold at a premium.

QUARTERSAWN BOARDS

Quartersawn boards are prized for their dimensional stability. They are more costly to produce.

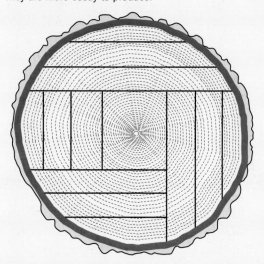

Sawing around a log is more time-consuming, but yields more premium quartersawn boards.

HOW WOOD DRIES

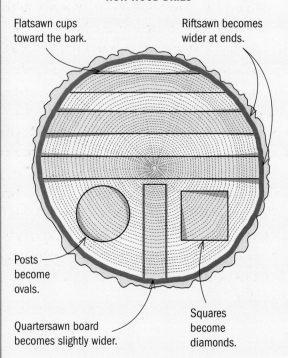

Flatsawn cups toward the bark.

Riftsawn becomes wider at ends.

Posts become ovals.

Quartersawn board becomes slightly wider.

Squares become diamonds.

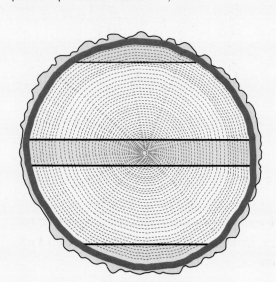

Flatsawing yields only a few quartersawn boards.

ask for 5/4 lumber, pronounced "five quarter." A roughsawn 5/4 board cleans up to 1" thick.

As if that weren't confusing enough, some places are now selling lumber from select species in which the boards are actually the dimensions on the label. When they say 1×4, they mean 1" thick and 4" wide.

The easiest way to get exactly what you want is to tell the yard staff the actual dimensions you need. If you ask for a "one by four" you could mean a board that's ¾" × 3½", or you could mean one that's 1" × 4". No one will be confused if you say, "I need a board that is actually 1" thick and 4" wide."

You can walk into the lumberyard with your nice tidy list, but chances are you won't be able to buy exactly what's on it. Say the list for your project specifies 4' 2×2s, but what if the yard has only 8-footers or 3-footers? Before you can make a choice, you must know the approximate final lengths of the pieces used in each project.

Some yards are self-service, where you're free to pick through the lumber to select the best boards on your own, but most often you'll be under the protection of a yard employee. Keep in mind that it's not customary for contractors to be overly fussy, and you don't want to be branded as that woodworker who obsesses about the perfect board. Be firm about rejecting the worst boards, but don't make the yard guys work too hard. It's better to keep their goodwill. You'll need it in the long run. Once you've put a salesperson through the hassle of shifting piles of wood so you can get a closer look, help stack it back as neatly as it was when you arrived. Or neater.

Whereas structural lumber is piled outside in the weather, **dimensional lumber is kept dry and out of the sun.**

time it's been kiln dried, cleaned up, and made straight it ends up as ¾" × 5½". Boards that start out as 1×6 or narrower end up ¼" thinner and ½" narrower. Boards wider than 6" end up ¾" narrower.

If you go into a lumberyard and ask for "one-by" lumber, the salesperson typically assumes you mean nominal 1" lumber, which is ¾" thick. If you want lumber that is 1" thick,

If you live near a woodworkers' store, a fine hardwoods dealer, or a local sawmill, you might choose to buy your wood roughsawn. The price will be lower, but you'll pay for the labor of milling and the wood it removes. A roughsawn board is harder to read, but at least you'll have expert assistance to help you pick out good ones.

Tell the staff what you need to end up with, and they'll help you select boards that will mill down to the correct dimensions. If you need a finished board that's 1" × 4", you may have to pay for a board that's 5/4 × 5½" or larger. You'll also pay an additional fee for the milling.

Whereas dimensioned lumber is typically sold as so many dollars per lineal foot, roughsawn lumber is sold by volume. It's an easy way to price lumber that's sold in irregular forms and of varying lengths, widths, and thicknesses. The unit is a board foot, which is equal to the amount of wood in a cube 1" × 12" × 12". Lumber dealers selling by the board foot needn't keep price lists for every possible combination of thickness and width. Three quick measurements are all you need.

To find the number of board feet in a piece of wood, multiply thickness, width, and length together (all in inches) and divide the result by 144. Multiply this figure by the price per board foot to get the cost of the board. This isn't easy to do in your head, but it's painless with a calculator.

Stacking Lumber While Conditioning

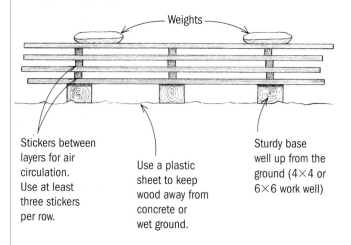

Weights

Stickers between layers for air circulation. Use at least three stickers per row.

Use a plastic sheet to keep wood away from concrete or wet ground.

Sturdy base well up from the ground (4×4 or 6×6 work well)

Storing Lumber

Once you've selected the wood, take it to your shop and let it sit for a couple of weeks so its moisture content can reach equilibrium with its new environment. Don't stack the boards one atop the other—it's important that air circulate around each one. Put stickers that are at least ¾" thick between each layer.

You can also store your lumber vertically against the wall, which allows you to select it for best figure, or in racks against the wall in your shop.

Sheet Goods

After all the trouble solid wood can cause, it's sometimes a relief to deal with stable, man-made materials such as plywood and MDF. Uniformly flat, dimensionally stable, easy to machine, and virtually unaffected by changes in relative humidity, sheet goods can solve all kinds of woodworking problems.

Plywood

Plywood is made from thin pieces of wood (called veneers) layered with each ply perpendicular to those on either side of it. The veneers are well glued together under high pressure and so thin that wood movement is impossible. Plywood is strong, waterproof, capable of bending around curves, and perfect where great strength and dimensional stability are required (see the drawing on p. 130).

Plywoods are specified by two different systems depending on whether they're hardwoods or softwoods. Softwood plywoods (such as pine and fir) are comparable to structural lumber and are too rough for most woodworking appli-

Sheet materials are dimensionally stable. **Shown are a high-end plywood with many plies (top), MDF (right), and hardboard (bottom).**

> **WORK SMART**
>
> **C**oncrete is like a sponge for moisture, so never put your wood directly on a concrete floor. Use stickers or plastic sheeting to keep the wood dry.

cations. Hardwood plywood is the woodworker's staple—it's used for shop jigs and fixtures, the backs and interiors of cabinets, and when made with handsome veneers, for furniture.

When buying hardwood plywood, first specify the quality of the faces, using letters:

- A faces use the best veneers matched for grain pattern and color.
- B faces are clear and matched for color but not grain pattern.
- C faces are not matched and may have plugs.

- D is a paint-grade finish that may have filled splits and holes.
- MDO face is covered with smooth paper; also called signboard because it paints so well.

Then specify the quality of the back:

- 1 is the highest quality, but it is not comparable to an A front.
- 2 has some wooden repairs.
- 3 has more repairs, sometimes filled with putty rather than wood and allows splits.
- 4 may have open splits and repairs that do not affect the strength of the panel.

Here's how it works. Let's say you want good-quality ¾"-thick plywood for making jigs, fixtures, and other general shop applications. You're not too concerned with appearance, but you don't want ugly scrapwood, either. Ask for ¾" B2 birch. Or you're going to build a lateral file and you want a handsome cherry veneer on the outside and a pretty good one on the inside. Ask for ¾" A2 cherry.

Although the outside dimensions of a sheet of plywood are a true 4' × 8', the specified thickness is nominal. The plies are laid up to yield, say, a ¾"-thick panel. After gluing it up at high pressure and heat, the panel is sanded flat. By the time both sides have been sanded, the ¾" has become ²³⁄₃₂" thick.

Plywood Construction

Plywood is dimensionally stable because the plies are oriented with alternating grain direction.

Plywood usually has an odd number of plies. More plies = higher strength and more uniform bending.

MDF

MDF stands for medium-density fiberboard, a homogenous material made from sawmill waste. Although it sounds a little like particleboard, MDF is a first-rate material. Sawdust, chips, and other wood debris are refined to a very small particle size, mixed with adhesive, and formed into sheets under extreme pressure. MDF machines beautifully, although cutting it makes a lot of dust. You can saw, plane, rout, and shape MDF, and it's easy to paint. It is more dimensionally stable than plywood, and unlike plywood, a ¾"-thick MDF panel is truly ¾" thick.

MDF has only two drawbacks to keep in mind. First, screws driven in its edges tend to split it, and second, it's heavy. A full sheet of ½" MDF is a handful, but some yards carry the lighter-colored, lighter-density version.

I like to use MDF around the shop for jigs and fixtures because of its even texture, good looks, and low price. A sheet of MDF costs

Be sure you know the final dimensions of the various pieces you'll be cutting so the yard can cut the sheet in a way that's useful to you. Remember that you can't count on yard saws to be accurate enough to cut to a final dimension.

about half as much as a sheet of plywood. Some yards offer "MDF lumber" in one-by sizes, primed and ready to paint. This is a great way to keep some ready-to-use MDF on hand.

Hardboard

You're probably familiar with hardboard in its pegboard incarnation—it also comes in a smooth-faced version. Similar to MDF in construction, hardboard is even denser and heavier. It's sold only in ⅛" and ¼" thicknesses. It's good to keep some ¼"-thick hardboard on hand because it's good for drawer bottoms, inserts, and for building things around the shop.

Beadboard

Take plywood or MDF and rout a series of lengthwise groove-and-bead profiles, and you have a stable, good-looking panel for walls, the backs of furniture, and cabinet doors. In this book, it's used for the back of the bookcase (see the Classic Bookcase project on p. 202). I prefer MDF beadboard because it is easier to paint and often comes primed.

Where to Buy Sheet Goods

You can usually get birch, luan, and oak plywood at home centers; lumberyards will likely have a wider variety of veneers, including cherry, maple, and luan. For rarer veneers, find a plywood specialty company that is accustomed to dealing with cabinet shops (see Resources on p. 296). MDF is readily available in lumberyards and home centers, and most places carry some type of beadboard.

You'll typically have to buy a full sheet—although some dealers offer half and even quarter sheets of their best sellers. If you must buy a full sheet, check to see if the yard is willing to make a cut or two to help you get it into your vehicle.

Fasteners

You can't always use glue—maybe you're in a hurry or you may need to be able to take a joint apart. You need to know a little about metal fasteners to hold pieces of wood together.

Wood Screws

Screws are the most useful fasteners in the woodworker's arsenal. They're stronger than nails, can pull a joint tight on their own without clamps, are removable, can be driven at almost any angle, and can be rendered invisible by using plugs (with clear finishes) or fillers (with paints).

The key to the screw's strength and usefulness lies in its threads, which wind around the root in a helix. As the screw turns, the helix bites into the wood, drawing the fastener through the piece of wood to be held and into the anchoring piece until the head of the screw is tight against the top piece, holding it in place. Within the anchoring piece, the threads cut through or crush some wood fibers, but most are simply shoved aside and end up molding themselves around the threads.

Dense hardwoods don't compress enough to allow the threads to mold, and the wood's resistance to the screw's passage can be enough to break the screw. If the screw doesn't break, something else must give, and the wood splits rather than allowing the thread to enter.

To prevent splitting, bore a pilot hole to remove some of the material in the anchoring piece to relieve the pressure. Drill the top part of the pilot hole at the shank wider so no threads engage the top piece and try to force it away from the piece below. Use a tapered countersink drill bit.

There are two basic types of wood screws on the market today: the traditional (made from a 500-year-old design) and modern production screws (see the photo at top left on the facing page). The design of a traditional screw stems from having once been cut by hand. The shank is the same diameter as the outside of

Ideal Proportions of a Wood-Screwed Joint

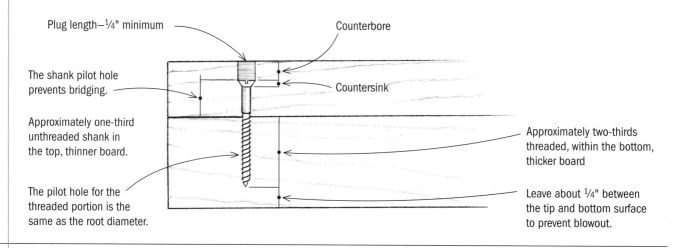

Plug length—1/4" minimum

Counterbore

The shank pilot hole prevents bridging.

Countersink

Approximately one-third unthreaded shank in the top, thinner board.

Approximately two-thirds threaded, within the bottom, thicker board

The pilot hole for the threaded portion is the same as the root diameter.

Leave about 1/4" between the tip and bottom surface to prevent blowout.

Two #8 × 1¼" wood screws: an old-fashioned screw (left) and a modern rolled-thread wood screw (right).

Lag screws are like course, long wood screws installed with a wrench. Always use a washer under the head to keep the wood around it from being crushed.

the threads, tapering slightly at the tip. The upper third of the screw is not threaded.

There are several proprietary production screw designs, and most have added a notch or extra thread or some similar feature at the tip that makes the screw into its own little pilot-hole bit. Except for very long screws, very hard woods, or fastening near the ends, you won't need to bore a pilot hole for these self-drilling screws. In addition, modern screws have aggressive thread designs that make them easier to drive by power.

I try to buy square-drive screws whenever possible. They're available with a countersinking head or with a wide washer head for pocket-hole screws or other applications where the load is spread over a greater area. The surface area between the bit and the screw is greater than with Phillips-head screws, and the bit stays in the slot at high torque. With a Phillips bit, high levels of torque often cause the bit to rise out of the slot enough to chew up the head as it rotates. You have to exert enormous pressure on a Phillips-head bit to keep it engaged. A square-drive screw is much easier to drive.

Lag Screws

A cross between a screw and a bolt, lag screws have large diameters, coarse threads, and big hex heads for driving with a wrench. They're the perfect solution for heavy-duty applications where the fastener is longer than about 3".

Lags need shank holes and pilot holes, just like traditional wood screws. Like bolts, lags

When woodworkers use nails, the nails are typically thin, small brads, or finishing nails such as these.

also need a washer under the head to spread out the force so the head won't sink into the wood, crushing the wood fibers around it (see photo at top right).

Brads and Finishing Nails

If a woodworking project calls for nails, it'll likely be for small, thin finishing nails or brads. Similar in size and proportion, brads and nails have different characteristics.

Finishing nails have sharp points and slightly larger heads, whereas brads have smaller heads that are easier to conceal. Finishing nails tend to be used by home builders. Brads are a hardware store item. Either is appropriate for the projects in this book—simply use what is available to you.

Glues

Strength is seldom a factor with modern woodworking glues; they're all stronger than the wood they bond. Other properties are far more important, like whether they're water-resistant, fill gaps, or have a fast cure time. Although there are hundreds of glues, most woodworkers stick with just a few.

Yellow Glue

The woodworkers' staple is yellow glue because it's easy to use and cures fast enough to remove the clamps in about an hour. Yellow glue is applied to both surfaces of the wood in an even coat just thick enough to obscure the grain pattern in the wood. A glue joint with the right amount of glue will squeeze out an even line of beads its whole length when clamped tightly. If there are no beads, you may have a glue-starved joint. On the other hand, if the glue runs out of the joint, you've used too much (see the photo below).

Once you've applied the glue, you've got 10 minutes to 15 minutes of open time, depending on the temperature and humidity. After that, the surface of the glue has started to cure, and you can't count on getting a good joint. Some companies make extended-open-time yellow glue, which can be a big help for complicated glue-ups and on hot days.

Yellow glue has no gap-filling qualities; for full strength the joints must fit very well and

If the glue runs out like this you've got too much. **Don't try to clean it up—you'll just spread it around. Wait until it's dry and scrape it off.**

the clamps have to be cranked tight. Because yellow glue is a water-based product, it causes the wood to swell immediately adjacent to the joint. This is important to remember, for if you plane or sand the joint flat too soon after gluing, when moisture evaporates, the wood around the joint will shrink into a low spot around the joint.

Yellow glue cleans up easily with water, but be careful to remove all traces. A thin film left on the surface may be invisible but will seal the wood like a finish, preventing stain from penetrating. If you've applied the right amount of glue, there's not enough squeeze-out to be a problem. Just leave the beads of glue to dry, then scrape them off cleanly.

For gluing dark woods or for highlighting gluelines, try dark-tinted yellow glue. Although light brown in color when applied, it dries almost black. For applying moldings or working on vertical surfaces, you can get thickened yellow glue that is less runny. Outdoor projects require the water-resistant version of yellow glue.

Epoxy

Epoxy is the only glue that bridges gaps and makes poorly fitted joints as strong as perfect-fitting joints. Since epoxy dries clear, it's also great for filling flaws in wood, like wormholes and knots. Just fill the hole with a little epoxy (you can mix in sawdust, too, if you like) and sand it flush once fully cured.

Epoxies are two-part products that must be well mixed in the proper ratio or they don't cure. Exactly what the ratio should be between parts A and B depends on the formulation of your epoxy. For most home-use products, the ratio is simply 1:1, with a margin of error of about 10%.

Open time and cure time depend on the formulation and can be anywhere from 5 minutes

WORK SMART

When cleaning up small drips and runs, don't just wipe with a paper towel. You risk spreading a thin film of glue over a larger surface. Scrub with a wet synthetic steel wool pad, then wipe up with a fresh paper towel moistened with clean water.

to 8 hours at room temperature. Epoxy cures much faster in warm weather—the cure time halves for each 10°F increase.

Uncured epoxy can be cleaned up with vinegar, denatured alcohol, or acetone. Once epoxy has cured, it's waterproof and impervious to pretty much everything except solar radiation.

Polyurethane

Honeylike in appearance, polyurethane glue cures in the presence of moisture. At room temperature you have about 20 minutes of open time, and the joints should remain in clamps for about 2 hours. You can speed the cure by moistening one surface with water before clamping. Polyurethanes are unique in that the squeeze-out turns to foam, which is easy to remove from the surface.

Polyurethanes are not gap-filling adhesives, but they are waterproof and can be stained.

Cyanoacrylate

Cyanoacrylate is the woodworker's version of Super Glue®. It cures very fast (30 seconds to 5 minutes depending on formulation) and is crystal clear, making it perfect for fixing splits, failed plugs, small holes, and other minor problems. It is not suitable for large jobs, but keep a bottle around. Sometimes it can save the day by fixing a split or gluing back a busted corner.

Abrasives

Abrasives are used in woodworking for shaping, smoothing, and finishing wood as well as for sharpening plane irons and chisels. Rough sandpaper with a low grit number is used for shaping and flattening or any time you want to remove material quickly (such as when getting rid of old finish or flattening the back of a plane iron). Once the surface is flat (or rounded if that's your goal), you switch to a finer-grit paper to smooth away the scratches left by shaping.

Getting a smooth surface is simply a matter of replacing the scratches left by the coarse sandpaper with finer ones. Sand the surface thoroughly (by hand or with power), brush it off to remove the dust and abrasive particles, and switch to the next finest grit. Repeat until the surface is as smooth as you want.

Garnet paper is harder to find, but many woodworkers prize it because grit for grit it leaves a smoother surface than aluminum oxide. Garnet is a softer abrasive that tends to burnish the surface rather than cut it, leaving a lovely sheen. That burnishing can also be used to advantage when sanding woods like pine that tend to absorb stains unevenly. A surface sanded with garnet paper is less prone to blotching. It is available in sheets.

Sanding Bare Wood

Working wood, you'll use sandpaper ranging in grit from 36# (aggressive removal of material or finish) up to about 220#. Any finer grit than that and the surface becomes too smooth for stain to penetrate or for glue or a finish to adhere to.

Of all the minerals used in making sandpapers, only two are suitable for sanding bare wood—aluminum oxide and garnet. It's an aggressive paper; it cuts fast and rather harshly. It's good all-around sandpaper for woodworking and is readily available in sheets and disks.

Sanding Paint and Other Finishes

Finishes are harder to sand than bare wood and call for a more aggressive abrasive like silicon carbide. You'll use it in papers with grits ranging from 100# to 220# between coats and up to 2,000# for polishing.

You can get silicon carbide papers with special backing made for wet sanding. Once a couple of coats of finish are on and dry, many people prefer to sand with water or oil as a lubricant. Wet sanding is fast and efficient but best of all, it's dust free.

Pick the right abrasive paper for the job: **(left to right)** silicon carbide, garnet, aluminum oxide, wet-sanding paper.

Synthetic Steel Wool

You've probably used these pads around the house—you can get synthetic steel wool scrubbers with a handle and sponges backed with a thin layer of it. Unlike steel wool, these pads won't shed little metal shards, and they're great for wet sanding.

The pads are long lasting and reusable, so get one of each color. Green is the coarsest, which makes it best suited for rough work—you'll use this the least. Maroon is great to use between finish coats when wet sanding. The gray pads are good for intermediate coats when using oils, and the white is the finest. Use the white ones for final coats of oil and for applying wax and oil/wax finishes.

Stains

You can use stains to change the color of wood, but they won't make one species look like another. Although it's unreasonable to think that you'll ever get pine to look like cherry, it's not unreasonable to try to get the color of your pine piece in the same tonal range as a cherry piece in the same room. Think color, not species. This frees you up to experiment with mixtures of color that yield surprising results, like the rich golden brown you get from putting a wash of lemon yellow beneath a brown mahogany stain or the surprising results you get from a light wash of black on figured maple.

The range of effects is virtually limitless, and each finisher's taste so personal that I'm not even going to try to suggest color combinations. What you need to know are the qualities of the basic types of stains and how to apply them.

It's important to remember that staining merely colors the wood—it does not protect it. For that you need a clear top coat over the stain. Make sure the stain is completely dry before applying the top coat.

Prestain Conditioners

Many common furniture woods are difficult to stain. Pine, cherry, and birch are notable for their tendency to blotch because of variations in the density of the wood. The porous areas take on more color. To even out absorption, you need to partially seal the porous areas so they take up less stain. Most stain manufacturers sell a compatible product for this purpose, but

OIL- VS. WATER-BASED FINISHES

Oil-based finishes go on smoothly and cure when the volatile petrochemical solvents evaporate, leaving behind a tough film. This process typically takes about 8 hours, and as a result, the surface settles and bubbles and brush marks usually disappear. It's much easier to get a glassy-smooth surface with an oil-based finish. However, the solvents are flammable and in high concentrations are proven to have a negative impact on health over the long term. Consequently, chemists have developed a number of water-based finishes that cure by the evaporation of water.

Because of their chemistry, water-based finishes are on the whole less shiny and durable (especially outdoors) than are oil-based finishes.

The other drawback of water-based finishes is that they cure quickly, so brush marks don't have a chance to settle out. But they clean up with water, and the fumes are safe. This makes them ideal for finishing jobs done indoors (especially in the winter when you might be using a heater) and when you need to build up your coats quickly.

When you're finishing a project, it makes sense to choose either oil-based or water-based products and go with the same type and brand throughout the finishing process. By using a prestain conditioner, stain, and top coat of the same product line, you won't have any chemical incompatibilities that could cause your finish to fail long before its time.

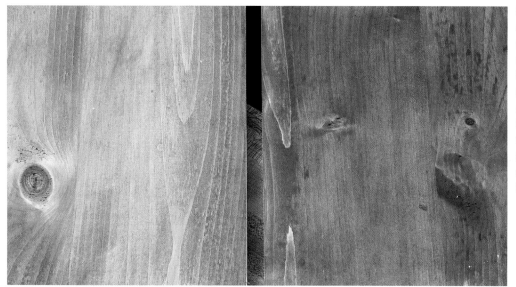

Some woods such as pine **soak up stain unevenly. Use a prestain conditioner to reduce blotchiness. The board shown at left got a light coat of thinned shellac before staining with a water-based pigment stain. The stain was applied directly to the board shown at right.**

I've had better luck using a thin coat of shellac, which works well under an oil- or water-based stain.

Pigment Stains

Made from colored powders (pigments) added to an oil or water base, pigment stains are commonly available in hardware stores and home centers. The pigments don't go into the wood as much as they sit on top of it, so pigment stains obscure the grain somewhat. If applied too thickly they look muddy, but they have the advantage of being reversible if treated with the proper solvent before drying. That means that if you don't like the effect you get with a pigment stain, you can remove most (if not all) of it with a solvent-soaked rag and try again.

Pigment stains settle in the open pores, accentuating the grain. Use them to highlight open-grained woods like oak, but be careful with woods like pine and birch. Their uneven grains are intensified and can look blotchy.

Dye Stains

Dye stains penetrate the wood and lend a more vivid, even color than do pigment stains. When applied, dye stains soak right into the wood, often leaving nothing behind to wipe up. To control color, thin the dye with the appropriate solvent (usually alcohol) and apply multiple coats.

Because of their even penetration, dye stains don't call attention to grain or figure in wood. That makes them your best choice for hard-to-stain woods such as birch and pine. Dye stains are available at specialty finishing and woodworking suppliers.

Gel Stains

These thickened, highly pigmented stains are a little easier to apply on vertical surfaces. They also offer more color control than standard pigment stains, making them a better choice on woods that tend to blotch. Some companies sell gel stains that are a combination pigment and dye stain and have the advantages of both.

Finishes

Most amateur woodworkers enjoy the process of finishing and consider it an important part of the process of building something by hand. You can get great results by learning to use only a few basic finishing methods.

Oil Finishes

The most common oils used in wood finishing are boiled linseed oil, tung oil, and Danish oil. All take a long time to dry but provide an inexpensive, moderately durable finish that's suitable for interior use. Oil deepens the color and tends to "pop" or highlight the figure in wood and has a soft, natural look. Apply with a rag, let the oil soak into the wood for a while, and wipe off the excess. Two or three coats are typically sufficient.

Oil finishes are easy to maintain—simply clean and "sand" the surface with synthetic steel wool and apply another coat.

Oil/Wax Finishes

The old-fashioned way to make an oil/wax finish is to melt beeswax into linseed oil, then thin with a little turpentine. These days the recipe includes modern fast-drying and durable polymers, but the soft sheen remains. This finish gives even new work an instant patina that is both rich and simple.

Apply the oil/wax finish in very thin coats with a rag or a piece of synthetic steel wool. Let it sit for 10 minutes to 15 minutes, and rub out with a soft cloth. It takes about 24 hours to dry completely. Two or three coats are adequate for protection, adding more deepens the patina. An oil/wax finish is suitable for pieces that won't be subjected to much heat or moisture.

Waxes

Wax is a soft but forgiving finish. Many people feel it's not durable to stand on its own, but I've lived for years with an antique French country dining table that's finished with wax and it still looks lovely. Hot plates and dishes can dull the surface, but all it takes is a little more wax applied with a shoe-shine brush to bring it back.

Wax goes well over any other finish—be it paint, varnish, shellac, or oil—and gives it a warm luster and silky feel. You can also use tinted waxes to impart some interesting effects. For instance, a dark brown wax over shellac or paint makes a piece look instantly old, especially when it lodges in knots, end grain, or surface imperfections.

Apply two or three thin coats of wax with synthetic steel wool and rub out with a soft cloth. If the surface is large, use a shoe-shine brush or a car polisher.

Shellacs

Shellac is made from the shell casing of female lac bugs (indigenous to India) dissolved in denatured alcohol. You can buy ready-made shellac in cans (be sure to get the more durable dewaxed variety), or you can buy it as flakes and make your own.

FINISHES AT A GLANCE

Type of Finish	Solvent	Durability	Best Uses	Maintenance
Oil (tung, boiled linseed, or Danish)	Turpentine or mineral spirits	Good indoors but won't last more than a few weeks outdoors	Any time you want to enhance the grain; for tables or other furniture that won't be used hard	Recoat periodically using synthetic steel wool.
Oil/wax	Turpentine	Good	Decorative items that aren't subjected to everyday wear and tear—boxes, display cabinets, figured tabletops	Recoat periodically using synthetic steel wool. Polish with a soft cloth.
Wax	Turpentine	Good	Tables and other furniture	Recoat periodically using synthetic steel wool. Polish with a soft cloth.
Shellac	Denatured alcohol	Fair	Use alone or as a sealer before or after staining or under an oil/wax finish for color.	Sand and recoat.
Varnish	Turpentine or mineral spirits	Excellent	Outdoor furniture or dining tables, suitable anywhere	Sand and recoat. When used outdoors, sand and apply two coats yearly.
Varnish/oil	Turpentine or mineral spirits	Excellent, but not as durable as straight varnish	Outdoor furniture or dining tables, suitable anywhere	Doesn't require sanding between coats; just make sure the surface is clean.
Water-based clear coats	Water	Very good	Tables, cabinets, etc.—anything indoors that gets hard use	Sand and recoat.
Latex paints	Water	Very good	Indoor and outdoor furniture	Sand and recoat (every few years outdoors).
Oil-based paints	Mineral spirits	Excellent	For a hard, shiny, durable surface	Sand and recoat every few years.

Shellac can be used as a finish in its own right or as a sealer coat beneath other finishes. It's the pro finisher's secret weapon because it solves myriad finishing problems. A very thin wash coat of shellac soaks into porous areas that tend to get blotchy when stained, lightly sealing the wood. Shellac reduces but does not eliminate stain penetration, effectively evening the color. A moderately thinned coat of shellac sticks to oily wood and prevents knots from bleeding resin so that subsequent finish coats stay put and look good. And finally, several coats of shellac on its own build up to a deep, rich finish with a warm glow.

Shellac is easy to apply with a brush or rag and dries quickly, allowing you to recoat in an hour or two. Since the solvent for shellac is alcohol, you needn't worry about toxic or highly flammable fumes. The only downside to shellac is that it is only moderately durable—and is certainly not the finish to use on a bar, since a spilled alcoholic beverage will act as a solvent. A coat or two of wax over shellac offers a bit of protection.

Varnishes

Varnishes are the toughest clear finish and the only ones suitable for hard service and exterior use. They're also the most difficult to apply. The relatively thick film takes a long time to dry and tends to pick up dust in the air. It will drip and wrinkle if not applied properly.

To apply varnish, use a foam brush. Thin the first coat 50% with the solvent suggested by the manufacturer. Thin the second coat 25% and all subsequent coats 10%. Although you can get by with as few as three coats, serious varnishers apply eight or more for a deep, rich finish that has no equal.

Varnishes are made with a variety of resins; you'll most likely have a choice between phenolic, alkyd, and polyurethane. The alkyd is the least durable; polyurethane is the most durable.

Gel Varnishes

A gel varnish is a slightly thickened varnish, typically polyurethane, that is meant to be applied with a rag. It's less messy than a regular varnish (which you can also apply with a rag) and slightly less durable. Apply thin coats.

Oil/Varnish Mixtures

Available in a variety of formulations with various degrees of sheen, pigment, and durability, oil/varnish mixtures combine the durability of varnish with oil's ease of use. Whereas you must sand varnish between coats, you can build up multiple coats of oil/varnish mixture without sanding.

Apply with a brush or rag and put on a few coats for a natural oiled look. Or you can apply several coats to approach the thick, rich look of varnish.

Water-Based Clear-Coat Finishes

You can't beat water-based finishes for ease of use. They dry quickly, emit no noxious fumes, and clean up with water. They're quite durable, too—only polyurethane varnishes are more resistant to daily wear and tear.

Perfectly clear, these finishes lack the rich glow of varnishes and oils, and many people feel they have a cold, plastic look. You can minimize that effect by applying water-based clear coats over shellac or by putting a few drops of amber tint (available at woodworking suppliers and paint stores) into the mixture.

Apply with a nylon-bristle brush and sand between coats.

Fine Paint Finishes

Paint is tougher and more durable than clear finishes and makes good sense for furniture used outdoors or in high-traffic areas. Paint also has a place in high-quality construction. A fine paint job is as handsome as natural wood.

The key to an awe-inspiring paint finish is to start with high-quality paint. You want one that's high in solids, with finely ground pigments and a hard, shiny surface.

It's easier to get a smooth painted finish with an oil-based paint, such as an alkyd or polyurethane enamel. Because of their longer drying times, oil-based paints self-level better than quick-drying water-based paints. Brush marks disappear as they dry. Lower-priced water-based paints, like common latex house paint, are designed to go on in a thick film and will almost always show brush marks. For painting furniture, look for an acrylic latex enamel or a waterborne polyurethane.

No paint looks good when applied over an ill-prepared surface. You've got to sand the surface smooth of all machine marks, make it flat and free of waves and ripples, and it must be smooth and free from rough grain, pinholes, dents, dings, and knots.

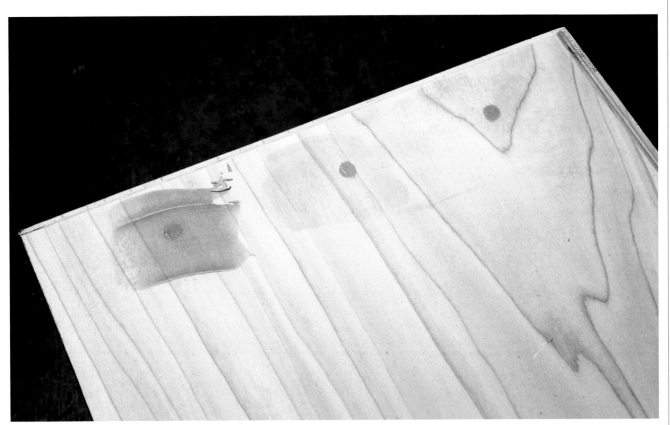

Properly sanded filler (right) shows a sharp delineation between the filled spot and surrounding wood. **The hole on the left hasn't been sanded at all; the one in the middle hasn't been sanded enough.**

So start your fine paint job when you sand your project, and work hard to keep the surface flat. Keep your sander moving and use hand-sanding blocks. As you sand the surface, keep your eyes open for surface imperfections and circle them with a pencil so you can fill them later with putty (see below).

Once the filler is dry, sand carefully until with your eyes closed, you can't tell when your fingertips are running over the putty.

A coat or two of primer under your top coat seals the surface and smooths out the wood grain. Look for a primer labeled "grain filling" or "high build." They're formulated to sand smooth in a short time and form a good chemical bond with paint. After priming, you may need to fill in a few places and sand again.

Finally, apply the paint in thin coats either by brush or from an aerosol spray can. The surface will end up smoother and dry more quickly than if the paint is caked on.

Putties and Fillers

Many beginning woodworkers make the mistake of thinking that wood putty can hide a multitude of sins. Small holes and cracks filled with putty tinted to match the wood end up pretty well camouflaged, but large flaws will remain evident.

If you're not using stain, you can make a thick paste of fine dust from hand-sanding and shellac or epoxy to fill the holes. These fillers don't stain well, but they are almost invisible under a clear finish. If you're staining the wood, use a putty that's near the wood's color before staining, and apply the stain to the wood and the filler. Even so, the repair will probably not stain in the same way as the wood.

If you're painting, you have a wider selection of fillers, since color is not important. The best fillers to use for a fine paint job come from an auto-body supply shop. Fill large areas with two-part polyester filler—it dries fast and is easy to sand. Once you've done the initial filling and are touching up pinholes, rough grain, and other small flaws, use an acrylic glazing putty (another auto-body product). These putties are light, smooth, easy to apply, dry quickly, and sand out beautifully.

Brushes

Four kinds of inexpensive brushes will serve most of your woodworking needs. Three of them are disposable. When using oil-based finishes, it's usually cheaper and easier to use disposable brushes than to buy, store, and properly dispose of the harmful solvents used to clean a higher-quality brush.

Acid Brushes

The coarse acid brushes are only ½" wide and make great applicators for glue. If you're using epoxy or polyurethane, just throw them away. But if you're laying down yellow glue, you can clean the brushes in soap and water and use them repeatedly.

Inexpensive Natural-Bristle Brushes

Natural-bristle brushes are pretty thin and they always lose bristles in the finish, but the price is right. They're not good enough to apply a film finish, but they're fine for applying stain that you're going to wipe off anyway or for an oil/varnish mixture. These brushes are available in sizes from 1" to 4".

Foam Brushes

Foam brushes produce a wonderful finish when used with oil-based paints and varnishes. Their smooth surface promotes self-leveling, and they don't hold too much finish to produce runs or drips. However, they're not a good choice for thicker water-based finishes.

Nylon-Bristle Brushes

Water-based finishes make natural-bristle brushes splay out and lose their shape, so you need to apply them with a synthetic-bristle brush. Look for a thick nylon brush with smooth, feathered ends. It's worth it to spend around $10 for a brush for water-based paints, since you can clean it with soap and water and use it again and again.

Other Finishing Supplies

In addition to paint and brushes, you need a few other items to achieve a fine painted finish.

Tack Cloths

Tack cloths are sticky swatches of cheesecloth that make it easy to get a smooth oil-based finish. After sanding and vacuuming the surface, wipe it down with a tack cloth to pick up any lingering dust and grit that could rough up your paint job. Be sure to buy compatible tack rags when using water-based finishes.

Paper Towels and Rags

There's always something to wipe up when finishing—if it's not part of the prescribed application process, then you're cleaning up. Most of the time I use a premium brand of paper towel for wiping up during finishing. Bargain brands are too flimsy and full of fine lint and dust. Just about the only time a rag is necessary is when rubbing out a wax finish—the friction is often too much for any paper towel. When rags are called for, they must be clean, lint-free cotton rags, which are sold by the pound in most home centers and paint stores.

Solvents

Whenever you buy a finishing product, make sure you buy the solvent recommended by the manufacturer. That's the only way to ensure it has the right chemistry to do what it's supposed to do. In addition, it's a good idea to keep a can of denatured alcohol on hand. It's a good all-around cleaning agent. You can use alcohol for cleaning off pencil marks, price tags, oil, uncured epoxy, and most unspecified goo. It's powerful but not dangerous.

To start woodworking, you need only four types of brushes: (bottom to top) disposable acid brush, foam brush, cheap bristle brush, and moderate-quality nylon-bristle brush.

Tapes

For most woodworkers, tape is a legitimate tool. Of course, every shop has a roll of duct tape, but it won't work in every situation.

Masking Tape

Aside from its proper purpose of protecting adjoining surfaces from unwanted paint, masking tape has a variety of uses around the shop. I use it to label parts (no pencil lines to sand off), to serve as light-duty clamps, to help hold clamping pads in place, and sometimes to protect surfaces from dirt and scrapes. I don't like the old-fashioned brown kind; it tears too easily when trying to get it off the roll or the project. I keep rolls of 1½" and ¾" bright blue tape at strategic locations in my shop.

Double-Sided Tape

Sometimes you can't beat double-sided tape as a light-duty clamp. It's the preferred way to hold a pair of workpieces together when routing or sawing, and you'll find other uses as well. For woodworking applications, you need tape that's at least ¾" wide. For added strength, choose a vinyl tape.

Outdoor Easy Chair

This is an outdoor easy chair with attitude. It evokes the sobriety of Shaker furniture, the honest simplicity of the Arts and Crafts movement, and the easy-living sloth of an Adirondack chair. Every line is straight and square, yet the chair feels funky and relaxed. Even though it's easy to construct from several small pieces of wood, it has a self-confident presence that works in almost any setting, from watching TV or playing video games in the family room or kid's bedroom to lounging by the pool with a book.

This funky chair has a lofty pedigree. It's based on the *Red Blue Chair*, designed by Gerrit T. Reitveld in 1917. Much has been written about it, from learned discussions of its form and meaning to studies of the spiritual significance of its geometry. The original chair garners high praise, even to the point of being called one of the pivotal designs in the history of architecture.

Reitveld, a cabinetmaker turned designer, was surprised by the praise. He had no intention of changing design sensibilities; he simply set out to design a good-looking and functional chair that could be made quickly and economically using simple joinery and off-the-shelf lumber.

Beginning woodworkers should take heart in the fact that such a simple premise could create a chair that turned the art world on its ear. It proves that building great furniture doesn't require a complex design or a master craftsman's skills.

What You'll Learn

- Measuring techniques that don't require a ruler
- Cutting multiples to exact length
- The essentials of glue-and-screw construction
- Using wooden plugs to cover screw heads
- Making and using simple jigs to increase speed and accuracy
- Working with epoxy
- Applying and maintaining an oil/varnish finish

Glue-and-screw joints **are easy to construct and make this chair redundantly strong.**

The Outdoor Easy Chair is an excellent project for learning the essential skills of getting and keeping things square. If your uprights vary in length, or if your horizontals aren't perfectly horizontal, you'll find that your chair wobbles. Perhaps worse, it'll lack the uplifting crispness that makes you want to sit down.

Since reading and remembering fractions is a source of trouble for many, I'll show you some simple tricks that can eliminate measuring. You'll learn how to use clamps and stop blocks to cut multiples to the exact same length, the essentials of glue-and-screw construction, how to make and use simple jigs to accurately locate joints, and how to apply and maintain an oil/varnish blend finish.

This chair features screwed joints reinforced with adhesive. The screws provide the joints' foundation of strength and act as clamps to hold the joints together while the adhesive dries (you can continue working without all those clamps in the way). This chair is best built with epoxy, a tough waterproof adhesive that's very forgiving. It bridges gaps between imperfect joints, making a powerful, water-tight connection that's stronger than the wood itself.

The versatile glue-and-screw joint is an important addition to your repertoire; it works well in a variety of situations. You can leave the screw heads visible for quick-and-dirty utilitarian construction (such as jigs), fill them with putty, or you can cover them with elegant wooden plugs set flush to the surface.

A strong joint is only part of what makes this chair last. Redundant strength is built into the design.

A simple jig made from leftover 2×2s (on the right) accurately positions the uprights (on the left) and holds them in place while driving the screws.

Outdoor Easy Chair

All base materials are 2×2 baluster stock (9 pieces of 4' long; 5 pieces of 8' long). Planks for back, seat, and arms are ¾" by 5½".

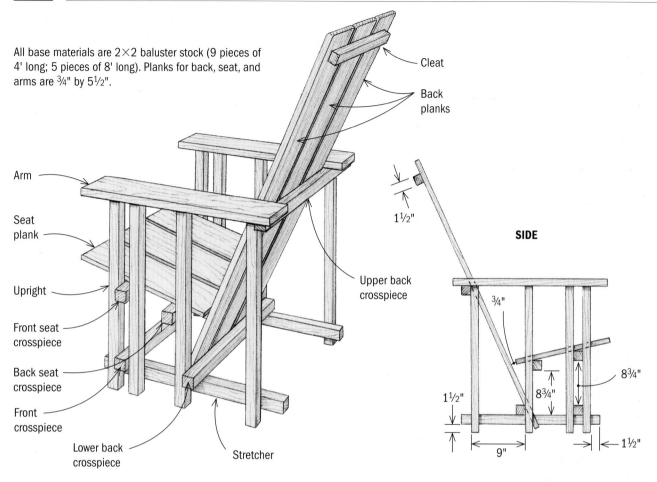

Cleat

Back planks

Arm

Seat plank

Upright

Front seat crosspiece

Back seat crosspiece

Front crosspiece

Lower back crosspiece

Stretcher

Upper back crosspiece

SIDE

1½"

¾"

8¾"

8¾"

1½"

9"

1½"

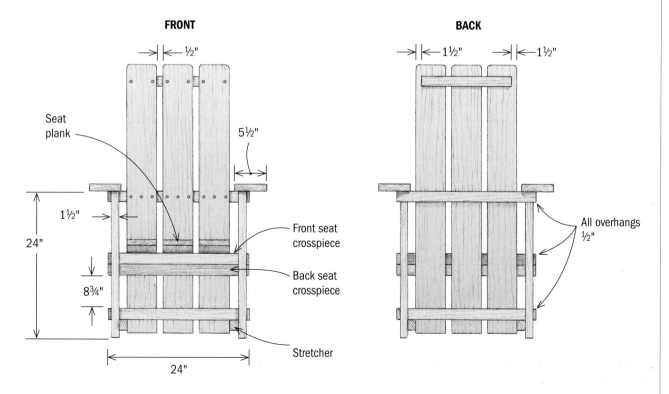

FRONT

½"

Seat plank

5½"

1½"

24"

8¾"

24"

Front seat crosspiece

Back seat crosspiece

Stretcher

BACK

1½"

1½"

All overhangs ½"

Building the Chair

When I'm teaching, I often put up a big sign that reads, "A pleasing result supersedes all measurements." I find myself pointing to the sign often. Most beginning woodworkers obsess about measurements, worrying far too much about whether their project exactly matches the specifications in the drawing. All that really matters is that parts go together square and true to one another. The actual final measurements don't matter.

To make this point forcefully, you won't do much measuring in this project. Instead of referring to a ruler, most of the time you'll use 2×2 offcuts to use as your measuring guide. This way, you can't focus on the minute markings on the tape measure or ruler; I want to encourage you to step back and look at the big picture. Even an untrained eye is surprisingly accurate about relative distances and can see if something is out of place. Before you drive any screws, see if it looks right.

The drawing on p. 151 shows dimensions, but your results may be different, since 2×2s can vary in width and thickness depending on what kind of wood they're made from. The actual dimensions are typically close to $1\frac{1}{2}$" × $1\frac{1}{2}$" but range to $1\frac{3}{8}$" and $1\frac{9}{16}$" (see pp. 127–128 for information on nominal vs. actual dimensions).

Before you start, remember to condition the wood by stickering it for a couple of weeks as described in the previous chapter. This is especially important for 2×2s sold for decking (meranti, ipé, some cedars). Decking lumber is often stored outside, giving it a higher moisture content. Without proper storage and conditioning, the wood might twist and bow during or

---
WORK SMART

Change your sandpaper often; when it gets dull and worn, it just takes longer to sand.

---

immediately after building. Better to condition the wood to your shop and discard any pieces that get too crooked in the process.

Build the Side Assemblies

Once your lumber has been conditioned, you're ready to get started. The first step is to crosscut all the 2×2s to the proper length.

Crosscutting and sanding base parts

1. Using the technique described in "Skill Builder: Crosscutting Multiples to the Same Length" on p. 156, cut the 2×2s into 17 identical pieces about 24" long (you need 14 for the chair and the other three are extras for mistakes and jigs).

If you're using a miter saw, your pieces will end up just a hair less than 24". A chopsaw has a thicker blade and makes a wider kerf; your pieces will be more like $23\frac{15}{16}$". It doesn't matter exactly how long the pieces are as long as they're identical.

Don't assume you can cut all your 4' 2×2s exactly in half with one cut. If they vary in length at all, the chair won't come out right. To be safe, position the stop block to cut a little less than 24". Cut the first piece to length, and

When you go to a home center or lumberyard and ask for balusters and matching one-by material, you'll have a selection of woods to choose from, depending on the local market.

■ **Red cedar**—The lightest in weight, the least expensive, and the least strong of the choices. Tends to split but works very well outdoors unfinished. Light pink to brown color. Paints well.

■ **Redwood**—Great weathering characteristics, light, strong, fragrant. Light reddish brown color; when untreated, it weathers to a silvery gray. Expensive.

■ **Port Orford (or white) cedar**—Moderate cost, moderate strength, paints well, takes exterior stains well. Light golden color, also weathers to a silvery gray.

■ **Meranti**—Similar in appearance to mahogany, substantially heavier than red or white cedar. Moderate in cost, weight, and strength. Looks good with a clear finish and takes paint well.

Often sold as decking lumber, it's sometimes called mahogany, though it's not a true mahogany. Sometimes sold prefinished.

■ **Oak**—Hard, strong, heavy. Paints well. Looks good with clear finish. It quickly turns black when exposed to water, so it's not suitable for clear finish outdoors. Moderately expensive.

■ **Ipé (or Brazilian walnut)**—Dense tropical hardwood often used for outdoor decks and railings. Dark in color, very heavy, very strong. Can be left to weather, or looks great with a clear finish. Moderately expensive.

■ **Pine**—Soft, light in color, not very strong. Doesn't stain well, knots may bleed sap. Expensive.

■ **Poplar**—Moderately strong, easy to work. Sometimes has a greenish purple cast that doesn't look good with clear finish or stain, but it paints very well. Moderate price.

A

then put the other piece against the stop block and cut a sliver from it.

2. Sand all four sides of each piece using a random-orbit sander and 150-grit sandpaper (see photo A).

Laying out the stretchers

1. Start by crosscutting one of the extra (or one cut too short) 2×2s to get two identical pieces 8¾" long for measuring blocks to use instead of a tape measure for several "measuring" operations.

2. Lay two 24" pieces on the bench, and get the ends flush by pushing them against one of the measuring blocks (photo B on p. 154 shows the operation to flush five pieces). Clamp them

together to prevent shifting while laying out the positions of the uprights.

3. Holding a measuring block as shown in photo C, make a mark on the inside edge. Keep the block in position, and place the other block alongside it, toward the center of the stretchers. Make a mark along the inside edge of the block, as shown in photo C. Lift off that block, and make hatch marks between the two lines. This is where the back upright crosses the stretcher. Mark this end of the stretchers "Back."

4. Using a ruler, make a mark at the back edge of the third upright toward the middle of the stretcher 9" from the line that represents the back edge of the back upright (the first line you drew in the previous step). Using the measuring block to mark the overlap, make hatch marks and label it "#3."

5. Mark the location of the first upright back from the front of the stretcher, using the measuring block and making hatch marks as in step 2.

6. Next, mark the position of the second upright exactly one block-width inward from the first upright. To make the marks, put the block back in place over the marks made in step 5, and put the second block along the inside of the first. Hold it in place and mark the inside edge. Don't let go of it yet.

7. Lift the first block, leapfrog it over the second, and put it down snug against the inside of the second block, as shown in photo D. Let go of the leapfrogged block, hold this one firmly, and mark the inside edge. Then lift it off to put hatch marks under it.

Laying out the uprights

The uprights extend below the stretchers by the width of one measuring block.

1. Align four of the uprights flush at the ends and clamp them as you did with the stretchers. (It's a little easier to get everything aligned doing four at a time rather than all eight at once.)

2. Lay a measuring block flush with one end of the uprights and make a mark on the inside end of the block across all four uprights.

3. Without moving this measuring block, hold the other measuring block against it and mark along its edge. The area under this block is where the screws go through the upright and into the stretcher. Make hatch marks, and label each piece as "Upright."

Countersinking the uprights for plugs

A **countersink** is nothing more than a funnel-shaped hole that matches the contour of the underside of a screw head. The countersunk hole allows the screw to nestle down flush with the surface of the wood. If you countersink deeply enough, the hole is far enough below the surface to contain a tapered wooden plug. It's an elegant way to hide unsightly screw heads in fine work.

1. Start by hanging the end of the workpiece off the edge of the bench so you don't drill into the bench, as shown in photo E.

2. Drill and countersink just a little less than ¼" deep (about 7/32") in the area that you hatch-marked. Locate the holes so they're catty-corner

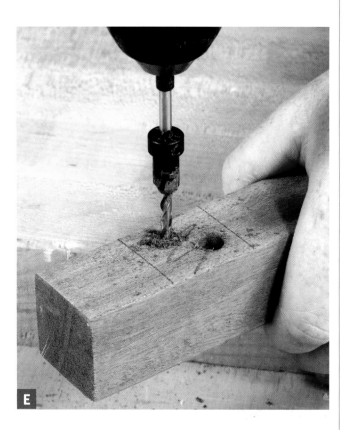

E

from each other using the methods shown in "Skill Builder: Countersinking for Screws and Plugs" on p. 158.

Fastening the uprights to the stretchers

To keep the chair from wobbling, the uprights must sit squarely on the floor, and the distance from the floor to the stretcher must be exactly the same on each. A simple jig made from a 2×2 makes it easy to position the uprights and hold them in place while fastening. (For more on this, see "Upright Jig" on p. 160.)

1. Lay one stretcher on the bench and clamp it down with the front to your right. This is important. The side assemblies are not interchangeable; they're mirror images of one another, as shown in photo F on p. 157. When you follow these steps for the other side, you'll place its front to the left. Put an extra piece of 2×2 about 20" above the stretcher to support the tops of the uprights while you work.

What You'll Need

- Crosscut saw
- Clamps
- Small scrap blocks with at least one square end

Crosscutting one or two identical pieces is typically done using clamps or double-sided tape. Once you're cutting more than a few, it makes sense to use a stop block, as shown in photo A.

A stop block is simply a piece of wood with a square end clamped in place to position pieces for identical cuts. If the workpieces are longer than the saw's fence, rig up an extension table to support long pieces, as shown in the photo and drawing below. Whether built for a miter saw or a chopsaw, the principles are the same.

1. Measure the first piece and put a mark where you want to make the cut. Square it across the face you intend to cut, as shown in photo B.

2. Clamp the piece in place on the saw table so the blade will cut on the line. Use the built-in hold-downs or small bar clamps.

3. Clamp a block to the fence so it's snug against the workpiece. Since you don't want the stop block to wriggle out of position during the course of many cuts, use two clamps.

4. Make the cut, remove the workpiece, and put another in place. Repeat until all are cut.

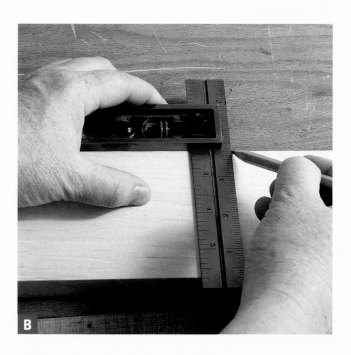

B

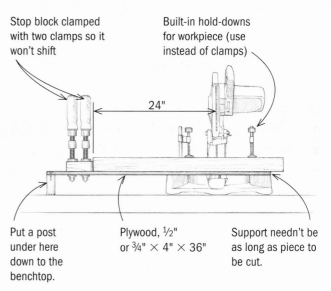

A

A Shop-Built Crosscut Extension Table

Stop block clamped with two clamps so it won't shift

Built-in hold-downs for workpiece (use instead of clamps)

24"

Put a post under here down to the benchtop.

Plywood, 1/2" or 3/4" × 4" × 36"

Support needn't be as long as piece to be cut.

F

WORK SMART

If your 2×2s are narrower than 1½", 2½" screws will be too long. Use 2" screws instead.

G

2. Apply glue to the hatched surface on the stretcher as well as to the back side of the upright at the countersunk holes, as shown in photo G.

3. Position the front upright on top of the stretcher with the glued surfaces together. Use the jig to locate it properly, clamping as necessary. (See the "Upright Jig" on p. 160 for details.)

4. Drive two 2½" screws into the countersunk holes. Be careful not to drive the screw too deep—if necessary adjust the clutch to get it right. Watch the joint to make sure the screw goes in all the way and the two pieces draw together so the excess epoxy squeezes out.

5. Repeat steps 1 to 4 for the second, third, and back uprights.

6. Place the other stretcher on the bench with the front pointing toward your left. Fasten the uprights as described above.

Connect the Side Assemblies with the Lower Crosspieces

Once you've built the two side assemblies, it's time to join them together with the lower crosspieces. They connect the stretchers at the bottom of the structure.

Laying out the crosspieces

The two sides are fastened together with a total of five crosspieces. These crosspieces extend ½" beyond the face of the uprights, giving the chair some visual texture.

1. Align the ends of the five crosspieces flush and clamp them together. Use your sliding

What You'll Need

- Cordless drill/driver
- #8 countersink/counterbore (stop collar optional)
- Scrap wood: ¾" pine and ¾" oak, two or three pieces about 4" × 14"

Countersinking for a Screw

A countersink is made by slipping a cone-shaped cutter onto the shank of a drill bit. The cutter makes a cone-shaped hole at the top of the shank hole to allow the head of a flathead screw to lie flush with the surface. In harder materials such as oak or MDF, a drilled countersink is necessary. When you're driving a screw near the edge of a board, a countersunk pilot hole will help prevent splitting.

1. Position the countersink for a 1¼" screw. Line up the tips of the screw and the bit, and slide the countersink so the top of the V that cuts the countersink is at the same level as the head of the screw, as shown in photo A.

2. If your bit came with a stop collar, set the stop collar right where the cutter changes from a V to straight sided. If you don't have a depth collar, you can use a piece of masking tape or go by eye.

B

C

3. Chuck the bit into the drill and drill the hole. Make sure the drill is set to high gear, full torque, and high speed. Alignment relative to the surface is important when countersinking. If you're not perpendicular to the surface, the screw will sit at a slight angle and won't be flush, as shown in photo B.

Countersinking for a Plug

If you drill the countersink deep enough, you can fill the hole with a wooden plug. If you have a stop collar, slide it into position about ¼" from the top of the funnel. Otherwise, use a piece of tape.

1. Drill a couple of holes, then insert the plugs using a hammer. If the hole is the correct depth, the plug will go in until it's about ¹⁄₁₆" above the surface. This makes it easy to trim flush. Find the correct depth by trial and error, as shown in photo C.

2. Once the depth is correct, drill several holes you can fill with plugs as discussed in "Skill Builder: Installing and Trimming Plugs" on p. 170. Keep the stop collar in position for countersinking the screw holes in the chair.

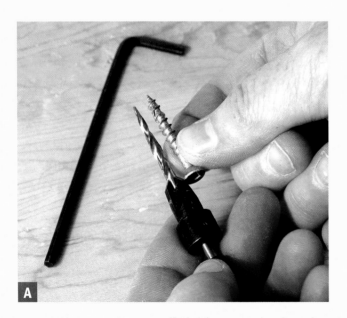

A

WORK SMART

If you make a mistake and screw the upright in the wrong place, immediately withdraw the screw, remove the upright, and fill the hole in the stretcher with epoxy. Once the epoxy has cured, sand it smooth and drive another screw in the correct location.

square to mark lines down both ends of all five that are ½" in from the ends.

2. Look at the crosspieces. Two of them fasten the stretchers together at a point two measuring blocks back from the line drawn in step 1. Let's call these crosspieces Type A. Two other crosspieces have their joints one measuring block back from the mark. These I'll call Type B. One crosspiece, Type C, needs no other mark. Use the blocks to "measure" out these distances as you did for the stretchers and upright, and make hatch marks in the joint areas, as shown in photo H.

Countersinking the crosspieces

Just as you drilled countersinks in the hatched areas on the uprights, do likewise in the hatched areas of four of the five crosspieces. The fifth crosspiece gets no holes at this time.

Connecting three crosspieces to one side

1. Put one side assembly on the bench so the stretcher sits on the top, with the extended bottom of the legs clearing the edge of the bench. Hold it in a vise or with clamps. Lay the scrap or extra crosspieces on the bench to support the far end of the crosspieces, as shown in photo I.

2. Select one Type A crosspiece to fasten between the two front uprights, positioned with the countersinks up. Apply glue to the bottom and both sides in the hatched area on the crosspiece and to the mating surfaces on the side assembly. Make sure the crosspiece extends beyond the upright by ½", as shown in photo J on p. 161. Clamp it in place while you drive two 2½" screws.

3. Do likewise with the lower back crosspiece (Type A), gluing and screwing it to the back edge of the third upright.

This simple jig takes all the fight out of locating the uprights (see photo A). It keeps the bottoms of the uprights in line and ensures perpendicularity. It also makes it easy to get the correct spacing between the first and second uprights.

Clamp the jig to the stretcher, as shown in photo B. In most cases you'll be able to drive the screws as shown, but you may have to hold the upright to the jig or even clamp them together.

A

Jig for Positioning Uprights

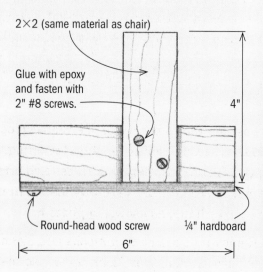

2×2 (same material as chair)

Glue with epoxy and fasten with 2" #8 screws.

4"

Round-head wood screw

¼" hardboard

6"

B

4. Lift the assembly out of the vise and place it on the benchtop, then clamp the other side assembly to the stretchers. Make sure everything fits, the ½" overhang is correct, and that the assembly is square. Then clamp the upper back crosspiece (Type B) in place.

5. Remove the crosspieces one at a time and apply epoxy to the mating surfaces on the crosspieces, stretchers, and uprights. Clamp in place once more (double-check the ½" overhang) and drive the screws. Then move on to the lower back crosspiece (Type A). Finally, fasten the upper back crosspiece (Type B).

Install the Seat Supports

The seat supports connect the side assemblies near the middle of the structure.

Installing the back seat support

1. Place a measuring block vertically on the stretcher, along the front side of the third upright. Use one per side and clamp them in place (only one clamp is necessary).

> **WORK SMART**
>
> **W**hen clamping the upper back crosspiece (Type B) after applying epoxy, position the clamps so that the pad covers one screw hole and leaves the other accessible. Once you've driven one screw per side, remove the clamps to install the other.

2. Position the remaining Type B crosspiece atop the measuring blocks while fastening. Apply epoxy to the mating surfaces and screw the crosspiece in place, carefully maintaining the ½" overhang, as shown in photo K.

3. Remove the measuring blocks and clean them with alcohol, then clean up any squeeze-out at the joint.

Installing the front seat support

1. Position a measuring block vertically atop the lower front crosspiece, between the first and second uprights. Clamp in place, as shown in photo L on p. 162.

2. Place the only Type C crosspiece atop the blocks. Clamp in place and countersink two

holes on each side through the front upright and into the crosspieces, as shown in photo M.

3. Remove the front seat crosspiece to clear away the dust from between the joint and apply epoxy to all mating surfaces.

4. Replace the front seat crosspiece, carefully aligning the ½" overhangs. This is hard to do without spreading a bit of epoxy around, but do your best to keep the mess to a minimum. Drive the screws, remove the blocks, and clean up.

Install the Back and Seat Planks

A ½" gap between the planks provides ventilation and prevents water and leaves from gathering in the seat. Install the center plank first, then the two sides using a piece of ½"-thick scrap.

Cutting the back and seat planks

1. Crosscut the three back planks from 1×6 stock to 48" long.

2. Crosscut the three seat planks from 1×6 stock to 17½" long.

3. Using a random-orbit sander and 150-grit disks, smooth all surfaces of the planks before assembly.

Installing the center back plank

If your workspace has low ceilings, you may have to move your chair to the floor for the following steps. If you have 50" above your workbench, keep the work on the bench and save your back.

1. Using a tape measure, find and mark the midpoints of the upper and lower back crosspieces as well as the front and rear seat crosspieces.

2. Using a ruler, mark a line half the width of a 1×6 (that's 2¾") out from the centerline on each side. The edges of the center planks will align with these marks.

3. Place two pieces of scrap 2×2 under the chair so the lower edge of the plank can rest on them. Use clamps to lightly hold the plank to the upper back crosspiece. Photo N shows the markings in step 2 as well as the extra pieces running fore and aft between the bottom of the plank and the bench.

4. On the back of the plank, draw lines along the upper and lower edges of the upper back support crosspiece, thus showing where the glue will go. Do the same on the lower back support crosspiece.

5. Next, determine the locations of the screws that will fasten the planks to the upper and lower back support crosspieces. Stand in front of the chair and bend or kneel so the upper back support crosspiece is at eye level. Draw a light horizontal line with a pencil on the plank along what appears to be the top of the crosspiece, as shown in photo O. You'll drill and countersink for screws just below along this line. Do the same with the lower back support (if your chair is on the floor, this will be awkward).

6. Remove the plank and use a square to draw perpendicular lines about ⅛" below the lines you drew by eye. This extra depth is insurance against the angled screw missing or coming out through the edge of the crosspieces.

7. Mark the locations of three holes: one in the center of the plank and a hole 1" in from each edge. Drill and countersink for screws on both the upper and lower crosspieces.

8. Apply epoxy to the mating surfaces and replace the center plank, using your lines to get it positioned. Clamp lightly. Drive the first screw in the upper right countersink and check the position of the plank. Is it still square? Did it shift? Reposition as necessary, then drive the second screw in the lower left countersink. Now the plank cannot shift and you can drive the remaining screws in any order.

WORK SMART

Don't crank down tightly or the clamps will skitter off the plank. Clamps are designed to hold things flat, and the planks cross the crosspiece at an angle. If you set the clamps loosely, they'll hold the plank in position well enough to do what you need to do.

Installing the side planks

1. Slide the scrap 2×2s from beneath the center plank so they'll support the bottom of the right plank. Place a ½" spacer about 12" long between the two planks near the middle of the back, then put a clamp across all three pieces—the two planks and the spacer. Slip another spacer at the bottom and clamp there, too. Make lines on the upper back crosspiece at each edge of the plank so you can quickly reposition it later, as shown in photo P.

2. Remove the plank, apply epoxy to the mating surfaces, reclamp, and fasten in place.

3. Now that you've had a bit of practice, you can save a step when installing the left side plank. Just apply epoxy, clamp in place with the spacers, and drive the fasteners.

> **WORK SMART**
>
> If you accidentally drill three holes in each outside plank, the outer two won't go into the cleat. Camouflage those holes by countersinking from both sides and filling each countersink with a plug.

Installing the seat planks

To install the seats, follow the same procedure as with the backs, but with one difference. Whereas you used the measuring blocks to hold the back planks off the floor, now you'll use a piece of ¾"-thick scrap as a spacer between the seat planks and the back planks.

1. Position the center plank on the marks 2¾" from the center with the ¾" spacer between the seat plank and the back.

2. Gauging by eye, draw a light pencil line along the center of the crosspiece. Also mark screw holes in the center and 1" from each edge.

3. Use the ½" spacers to position the side planks the correct distance from the center plank and the ¾" spacer to position them relative to the back.

4. Glue and fasten in the same manner as the back planks.

Install the Back Cleat

If you stand behind the chair and look down at the tops of the back planks, they probably don't lie in a nice straight line. That's not because of anything you did as a builder, but because the wood is not perfectly flat. It may have some bow along its length or some twist. It may even be cupped a little. For a crisp, refined appearance, fasten a cleat near the top of the back to hold the planks in alignment, as shown in photo Q.

1. Measure the width of the back planks, then crosscut a leftover 2×2 to a length that is two block widths less and sand it smooth using a random-orbit sander and 150-grit paper.

2. Clamp the cleat in place so it's one measuring block down from the top and one measuring block in from each end, as shown in photo R.

3. Draw around the edges of the cleat with a pencil to show where the glue will go. Unclamp

5. Using only the drill portion of your countersink bit, drill a pilot hole for each screw through the planks from back to front. Don't let the countersink cut the wood; just bore a hole, as shown in photo S.

6. Go around to the front of the chair, and using the pilot holes as guides, countersink proper holes for plugs.

7. Apply epoxy to the back of the cleat and to the rectangles on the back planks, and clamp it in place. If your seat back planks are much out of alignment, it may take your most powerful clamps to get them in place. Drive 1¼" screws from the front—through the planks and into the cleat.

Install the Arms

If you are both lucky and careful, all your uprights will be perfectly aligned. More likely, a couple of uprights are slightly bowed, or maybe one didn't get installed quite right. The arms are designed to camouflage such discrepancies.

1. Crosscut a 1×6 into two pieces, each 27" long, and sand the planks using 150-grit paper and a random-orbit sander.

2. Clamp the ½"-thick plank spacer to the inside of the uprights with clamps at the front and back. Make sure the spacer doesn't extend

the cleat to reveal the rectangle drawn on the back side of the back. The screws that hold the cleat to the back planks will be on the centerline of the rectangle. The center plank has three holes—one on center and one hole 1" in from each edge. The side planks have only two holes each—one on center and one 1" from the inside edge.

4. On the back of the planks, mark the locations of the screws that will hold the cleat in place.

above the tops of the uprights. Use another clamp to try and pull any misaligned uprights into place, but don't worry if they're not perfect. Photo T shows this clearly.

3. Mark the edges of each upright on the face of the spacer to show you where to drill later, when the arm will obscure your view up the uprights.

4. Set your sliding square so the base is on the spacer and the end of the blade is flush with the outside of the upright. Then set the arm on top of the uprights so that the inside edge is flush with the spacer and the back edge overhangs the upper back crosspiece by the width of one measuring block. Clamp front and back.

5. Even though you can't see the tops of the uprights, you can use your sliding square to mark their locations. The spacers already have marks on them showing the front and back sides of the uprights, and your square is now set to the outside edge. Mark these on the top of the arm, and reset your sliding square to ½" to get the inside edge. You don't need to be overly fussy about this—all you need is to locate the screw at or close to the center of the upright.

6. Mark the locations of the screws by eye in the middle of the square you've drawn to represent the upright. At the back of each arm, mark the locations of two countersinks for screws to go through the arms and into the upper back support crosspiece. Drill and countersink the holes, as shown in photo T.

T

7. Put epoxy on the tops of the uprights and around the holes on the underside of the arms (just make it about the same size as the upright by eye). This being end grain, most of the strength comes from the screws, but the epoxy does add a little extra strength and, more important, seals the end grain against absorbing moisture.

8. Drive a 2½" screw into each of the uprights and two 1¼" screws into the back crosspiece.

Your chair is now structurally sound; before you do anything else, have a seat. You've earned it!

Install the Plugs

You'll have more choices in wood for your chair than you'll have for the plugs. If you can't get the same species of wood for the plugs, consider a contrasting color. The chair shown is made from meranti with oak plugs. Try to get face-grain plugs—they're more stable and easier to cut than end-grain plugs.

Install all of the plugs in your chair according to the method described in "Skill Builder: Installing and Trimming Plugs" on p. 170, except that you'll put a little epoxy in the hole to keep the plugs in place. Allow the epoxy to fully cure before cutting the plugs.

Final Smoothing and Shaping

You did the lion's share of the sanding before assembly. Now all you have to do is clean up any errant epoxy, flatten the plugs, round the ends of the boards, and smooth the whole piece.

1. Lightly chamfer the ends of all the pieces with a block plane or sanding block. Pay special attention to the areas that come in contact with your body when you sit in the chair—the fronts of the arms and the front of the seat.

2. Also chamfer the ends of the back planks and arms, the ends of the crosspieces, and the ends of the stretchers. Cut light chamfers, slightly less than ⅛" wide, then smooth them over using 150-grit sandpaper and a hand-sanding block.

3. Give the bottoms of the uprights a slightly stronger chamfer (a hair more than ⅛") that's not rounded. This will make the chair appear lighter on its feet, almost as though it were floating.

4. Hand-sand the entire surface of the chair using 220-grit paper. Go with the grain, and pay attention to the areas around the joints where you might have glue squeeze-out and to the area around the plugs. You might need to use the chisel—bevel up—to remove a blob of epoxy.

Sealing the bottoms of the uprights

End grain soaks up water, which will tend to lift whatever finish you put on your chair. To prevent this, seal the bottom of the uprights with epoxy. It'll soak into the grain, so be prepared to put on a couple of thin coats to thoroughly seal it, as shown in photo U. (If yours will be an indoor chair, you can skip this step.)

V

WORK SMART

The amount of time you spend sanding, filing, and smoothing your project makes a big difference in the overall appearance and feel of the piece. Don't rush through the sanding; it's as important as the building.

Finishing

The best easy-to-maintain clear finish for exterior use is a mixture of natural oils and modern varnish resins. This finish also looks good on chairs intended to be used inside. I've chosen not to stain this project, but if you want to change the color of the wood, be sure to use a stain intended for exterior use (see the previous chapter for more information on stains and staining).

Oil finish alone is easy to apply and visually enhances the grain, but it doesn't hold up to ultraviolet light and can take days to dry.

Varnish provides a smooth, durable finish but requires 8 to 10 coats to look its best and can be fussy to apply. Mix them together, and you get the best of both worlds—a finish that looks great, is easy to apply, dries quickly, and is durable. (See p. 142 for more information on oil/varnish blends.)

1. Before applying the finish, repair minor divots, imperfectly fitting plugs, cracks, and other blemishes with epoxy. Simply fill a divot or crack with epoxy, or flow it into the space around an imperfectly drilled or cut plug. Cured epoxy dries clear, much like the finish, and the repairs will be hardly noticeable (see photo V). Sand it flush by hand.

2. After sanding, vacuum the chair and the surrounding area thoroughly to remove all sanding dust.

3. Wipe everything down with a tack cloth to get the last of the dust.

4. Apply a thin coat of oil/varnish mixture using an inexpensive natural-bristle or foam brush, as shown in photo W. You can apply a second coat after 8 hours.

5. When that coat is thoroughly dry, you can apply another coat or two without sanding. In theory you needn't sand between each coat of oil/varnish mixture, but in practice you'll get a smoother finish if you do. I handle this by apply-

ing two or three coats and then sanding before applying another two coats. If I want a really rich, smooth finish, I'll sand once more and apply another coat.

6. Once you've got some finish on your chair, you can eliminate all the hassles of dust by wet sanding. Rather than making dust, wet sanding makes a slurry that you simply wipe down or hose away. Just make sure you get the slurry off your chair before it dries, or it will be harder to remove. Wet-sand using a maroon synthetic steel wool pad.

7. Maintain your chair's finish by repeating step 6. If the chair lives outdoors, reapply once a year, twice a year in particularly sunny climes. If it's an indoor chair, you may go years before recoating.

What You'll Need

- ■ ³⁄₈" flat-grain plugs
  (also called face-grain plugs)

- ■ ³⁄₄" or 1" bench chisel

- ■ Small flat-faced mallet or hammer

- ■ Fast-cure epoxy
  (a self-mixing applicator is neat and easy)

- ■ 150-grit sandpaper
  (if sanding end-grain plugs, use 100-grit first)

- ■ Hand-sanding block

A

P ractice setting and trimming plugs in scrap wood before you start on your chair. And when you do plug your chair, begin with plugs in hidden locations until you hit your stride. Save the most visible plugs (in the back and arms) for last. Make sure your chisel's sharp before you begin.

1. Drill pilot holes and countersinks, then put a little glue in the hole and insert the plug. Align the grain of the plug with that of the surrounding wood. Set the plug with a small flat-headed mallet or hammer, as shown in photo A, but don't hit the plug too hard or it'll split. Be sure to listen: When the plug is snug against the screw head, the sound of the hammer blow changes to a clear, ringing thunk.

2. Once the glue has cured, cut the plug flush using a sharp chisel. It's a two-step process. First, use the hammer and the bevel of the chisel down to cut perpendicular to the grain to remove most of the plug, as shown in photo B. Then ditch the hammer and switch to paring with the bevel up.

3. Start the cut at the back of the plug. Don't try to cut much off the plug, never more than about ¹⁄₃₂". Tap lightly using the hammer, tilting the chisel to control depth as you go. Make very light cuts to produce paper-thin shavings, as shown in photo C. If you cut too deep, you'll lose control and possibly mar the surface.

4. Turn the chisel over and put the back flat on the wood. Press the back firmly with your hand to keep it flat on the surface, and move the chisel back and forth to pare away the last slivers, as shown in photo D.

5. Finally, sand the plugs smooth using 150-grit sandpaper and a random-orbit sander or a hand-sanding block. If you're using a machine, keep it moving and don't concentrate on the area around the plug.

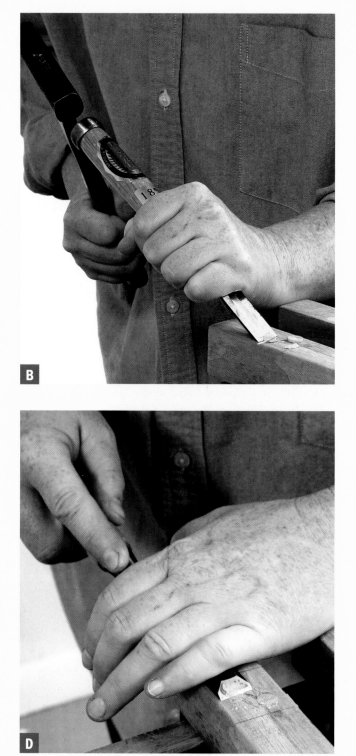

Old-World Coffee Table

This table speaks of a simpler life, of Old-World farmhouses, bountiful harvests, golden sunlight, and long Sunday afternoons with friends and family. Old-fashioned, sturdy, and unpretentious, it's a piece of furniture you can live with for years.

The ready-made legs, with their big curves and generous proportions, provide a firm foundation for the table. They're connected by wide aprons featuring a beaded bottom edge. The top is made from a panel glued up from four narrower boards; its edge is a simple and graceful roundover.

Coffee tables are one of the hardest-working pieces of furniture in the modern home, and this rustic-looking table is designed to handle all the rigors of modern life—yards of Super Bowl® snacks, a squirming child, impromptu dinner for the family, mountains of books, or a pack of roughhousing kids. It's engineered to be strong enough to handle the load, stiff enough to bear it without bending, and so well put together that a variety of stresses won't loosen its joints.

By the time you've finished this table, you'll have a solid grasp on some important woodworking skills: how to glue and clamp up boards to form a wide panel, how to make a decorative bead and use the tool to do it, how to round over an edge using a plane, and an effective method for preventing your stained finishes from looking blotchy. You're not only building a great coffee table but also increasing your skills.

What You'll Learn

- Making a pocket-hole screw joint
- Making and using a hand-beading tool
- Gluing up wide panels
- Using a circular saw and fence for accurate cuts
- Rounding corners with a plane
- Using an orbital sander for flattening and smoothing
- Getting a blotch-free stain on pine
- Using colored waxes as a finish

A pocket-hole screw joint **is easy to make, self-squaring, and strong.**

There are good reasons for making a table like this early in your woodworking career. Its fashionable rusticity transforms "problems" or "mistakes" into charm. The top needn't be absolutely flat; in fact, it won't look right if it is. The big legs draw the eye away from small problems and the offset at the joints hides any misalignments. Dents or dings that might occur during construction just give it character, and if the bead isn't straight and true, it's just proof the table was made by hand. No matter how the table comes out, once it's in your home, you'll see that the outcome needn't be perfect for the result to be wonderful.

Although the pocket-hole screw joint has been around for a long time, it is enjoying a new surge in popularity. Suddenly, every woodworking-supply outlet offers at least a couple of kinds of angled jigs for making the joints efficiently and accurately. It's an ideal method for fastening corner joints where end grain meets side grain, as between the aprons and legs of this table, or between the rails and stiles of a face frame.

But pocket-hole joinery is more versatile than that. You can also use it to fasten the corners of boxes, cabinets, and drawers (just locate the screws so they don't show), to fasten edging, and even to make curves. Building this coffee table will get you started in pocket-hole joinery; once you know how it works, you'll find plenty of applications on your own.

Long before there were routers and other machines to mold edges, woodworkers used simple details to add visual appeal to a design. The beaded edge is a perfect example—fun and easy to make, it livens up an otherwise bland edge. In this project, you'll learn to make and use a beading tool, thus adding to both your woodworking and your design skills (see photo A on p. 177). You'll also learn more about planes; they're not just for making things flat. In building this table, you'll see how you can use a plane to make a rounded edge.

Gluing narrow boards into wide panels is a critical skill you'll use again and again in your

Old-World Coffee Table

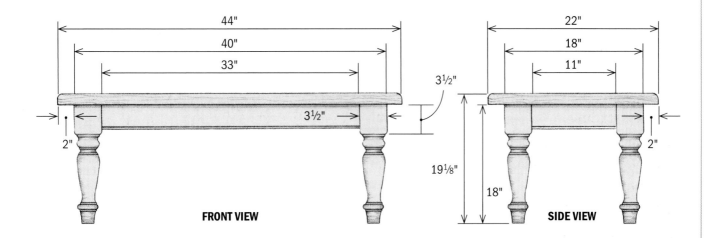

44"

40"

33"

3½"

3½"

3½"

2"

FRONT VIEW

22"

18"

11"

3½"

19⅛"

18"

2"

SIDE VIEW

CONSTRUCTION DETAIL

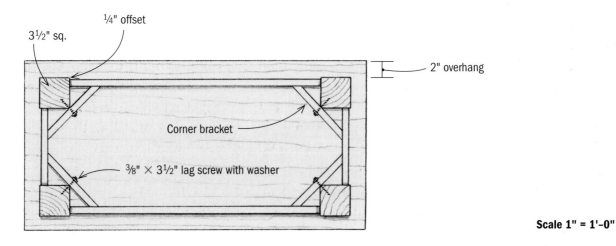

¼" offset

3½" sq.

2" overhang

Corner bracket

⅜" × 3½" lag screw with washer

Scale 1" = 1'–0"

projects, and you'll learn how to do it right. Using the methods and tools described here, you'll learn to get flat panels with a minimum of trouble.

The skills you learn in building this simple project apply to all kinds of tables, from teacup-size end tables to dining tables that seat 12. Your newfound skills in making panels and putting together pocket-hole screwed joints can be put to use making everything from jigs and shop furniture to kitchen cabinets and beds.

Quantity	Description	Actual Dimension*	Length	Notes
2	Pine lumber	1⅛" × 5½"	8'	Sold as 5/4 × 6". B or better select pine. If you choose #2 common, you may need 3 to yield 4 good boards each 44" long.
2	Pine lumber	1⅛" × 3½"	8'	Sold as 5/4 × 4" #2 common pine. Longest piece 33"
4	Jumbo English country legs		18"	Pine From Osborne Wood Products, Inc. See Resources on p. 296.
3	Hardboard scraps to shim offsets	¼" × about 3"	About 3"	Use offcuts from shopmade circular saw guide.
8	Figure 8 table clips			
4	Lag screws	⅜"	3½"	
4	Flat washers	⅜"		
16 (approx.)	#8 washer-head pocket-hole screws, square drive		2½"	
16 (approx.)	#8 washer-head pocket-hole screws, square drive		1¼"	
1	#10 steel flat-head wood screw		1½"	For beading tool
2 each	80-, 120-, 180-, 220-grit sanding disks			
2 sheets	220-grit sandpaper			
1 sheet	320-grit sandpaper			
	Glue			
	Dewaxed shellac			
	Denatured alcohol			For thinning shellac
	Water-based stain			
	Inexpensive natural-bristle brush			
	Dark-tinted wax			

TOOLS

- Tape measure
- Crosscut saw
- Beading tool (shopmade)
- Cordless drill/driver
- Pocket-hole jig
- ⅜" step drill for pocket holes
- 6"-long #2 square-drive bit
- Flexible shaft (optional)
- 1½" #2 square-drive bit
- #2 square-drive hand driver
- Straight screwdriver or Phillips screwdriver (depends on screws)
- Five panel clamps
- Six bar clamps
- Scraper
- Socket or other wrench to fit lag screws
- Block plane or bench plane
- 12" combination square
- Circular saw
- Saw guide (shopmade)
- ⅛" brad-point bit
- ⁵⁄₆₄" drill bit
- Random-orbit sander
- ¾" Forstner drill bit

Building the Coffee Table

M ake the aprons and build the support structure for your table, then make the top to fit.

Make the Aprons

1. Start by going through the $1\frac{1}{8}" \times 3\frac{1}{2}"$ apron material and figuring out how to cut it to avoid major defects such as ugly knots, knots at the end of an apron, damaged edges, and swirling grain. Lay out the boards so you can comfortably ponder the best use of material. Figure out where each piece should be cut, label the pieces, then crosscut them an inch or so too long.

A simple shopmade beading tool **makes a decorative detail on the lower edge of the aprons.**

2. Remember that the two short end aprons must be identical in length, as must the two long side aprons. If not, the corners won't go together at 90°. Clamp or tape them in pairs so they come out alike, as shown in photo B. The actual length is not critical.

3. To get a firm idea of how your table goes together, put the four legs upside down on the bench and lay the approximate-length aprons in place. Check the legs for knots and keep any big ones out of the way of the screws (knots are sometimes significantly harder than the surrounding wood). If there are any defects in the edges of the aprons, orient them so they won't get in the way of the beading and place them at the top of the base so they'll be under the tabletop where you can't see them (with the defect down on the benchtop in this orientation), as shown in photo C on p. 179. Mark the pieces so you can reassemble them the same way later.

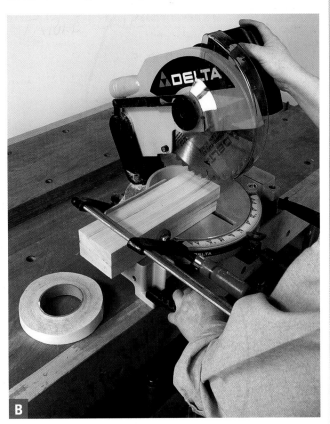

his basic design is all you need to know to fill your house with tables of all kinds. Just change the dimensions and most tables are within your capabilities.

For instance, a narrow end table with tapered legs is built the same way. It just needs narrower aprons and a smaller top. You can forgo the corner bracing on such a small table, as shown at right below. A dining table needs everything bigger—wider aprons, wider corner bracing, and depending on its length, a cross brace or two across the middle, as shown at left below.

Determining the size of the aprons is something you can readily do by eye. Size the apron so it looks good with the leg and it will be strong enough. Sketch the leg full size and simply try out a few widths until you hit upon something that looks right. Just remember to leave a little bit of the square part of the leg showing beneath the apron.

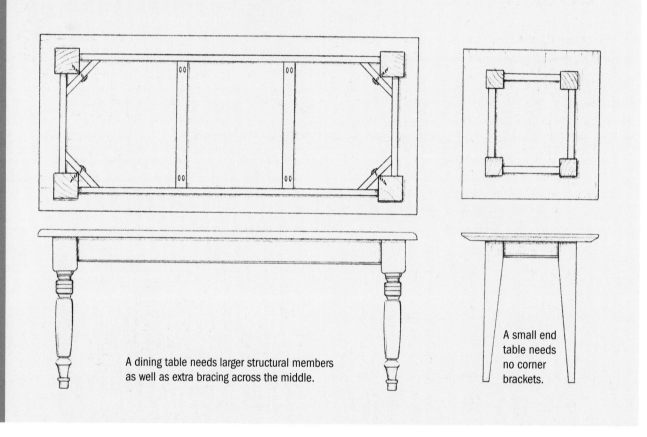

A dining table needs larger structural members as well as extra bracing across the middle.

A small end table needs no corner brackets.

Bead the Edge and Cut to Length

1. Use the beading tool and the method described in "Skill Builder: Making and Using a Beading Tool" on p. 180. Beading is similar to planing in that it's crucial the workpiece be held firmly, so use a vise or clamp the workpiece to the benchtop. If you use clamps, you'll have to figure out how to keep them out of the way. As the drawing on p. 181 shows, clamp thin pieces of wood tight up against the workpiece, using two clamps so nothing will shift.

2. Hold the tool as shown in photo A (on p. 177), and make several passes so the screw head slightly rounds over the wood, forming the bead.

Rustic furniture has been made of pine for centuries because it's light in weight and color, easy to work, and readily available. Most lumberyards and home centers have pine in all shapes and sizes, although the exact type of pine differs from region to region.

Unlike most other species of wood, dealers typically carry more than one grade of pine. Boards are graded by strength and appearance according to specifications laid down by lumber industry associations. Most lumberyards and home centers offer both common and select grades. Common grades

of pine (#1 through #5) are structurally sound but contain knots—#1 has the fewest, #5 has the most. The select grades (A through D) have fewer knots and a higher price. At my local yard, a B select board costs three times as much as a #2 common board with the same dimensions.

For the base of the table, #2 common gives an appropriate rusticity, but keep in mind that knots near the edges can cause boards to twist. Try to pick boards that are straight and don't have too many knots. For the top, you might want to go with select, but the highest

grades are too bland. I look for B select boards with a few pin knots to give character.

Make the beads on the edges where the grain is straightest and where there are no knots.

Do this for all four aprons. Don't press down too hard or go too quickly, or the bead will get wavy as the edge of the screw head tends to follow the grain (rather than remaining parallel to the edge).

Mark and Drill the Pocket Holes

1. Once more, set the legs and base upside down on the benchtop. Make sure all of the beads are oriented properly (up and facing out in this orientation). Clearly mark the

C

What You'll Need

- ■ 4"-long piece of scrap 2×2 or similar hardwood
- ■ One #10 × 1" flat-head wood screw (the old-fashioned kind, length is not critical, 1¼" is okay)
- ■ Cordless drill
- ■ Sandpaper
- ■ ⅛" brad-point bit
- ■ Hand screwdriver for the screw
- ■ Scrap wood for beading, 5/4 or ¾" (any width, at least 12" long)

A

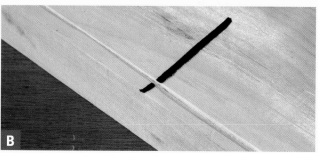

B

The decorative bead along the bottom edge of the aprons is distinctive and easy to make using a simple shopmade beading tool, as shown in photo A and the drawing at bottom right.

1. Find a nicely rounded piece of hardwood that fits in your hand (a 4"-long piece of scrap 2×2 from the Outdoor Easy Chair in the previous chapter is perfect). Sand it smooth.

2. Drill a ⅛" pilot hole in one face about 1½" in from the end and about ½" in from one edge. Drive a #10 steel wood screw into the hole until the underside of the head is about ¼" from the surface of the block.

3. Start the bead at whichever end is comfortable by pressing the block firmly against the side of the board and pulling the tool toward you. It's easier if you hold the opposite edge of the board with the other hand.

4. Run the tool the length of the board, pressing downward while pulling. Go slowly and be sure to keep the tool against the edge.

5. Take several passes, and stop when the underside of the screw starts rounding over the bead adjacent to the cut. Photo B shows only a few passes to the

A Shopmade Beading Tool

Cut good-looking rustic-style beads with this simple tool, which you can make from any scrap to fit your hand. The only dimension of importance is the ½" between the screw and the edge of the tool. Rounded corners are comfortable in the hand.

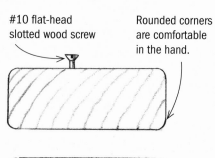

#10 flat-head slotted wood screw

Rounded corners are comfortable in the hand.

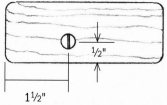

½"

1½"

left of the black line. More passes were made with the tool on the portion to the right of the line—the bead is more defined. You can adjust the size of the bead by changing the distance between the head of the screw and the surface of the block. A greater distance makes a wider bead.

Alternate Clamping Arrangement for Beading

If you don't have a vise to hold aprons for beading, you can use this arrangement to get the clamps out of the way. Be sure the pieces of scrap are clamped tight against the ends. Right-handers will prefer to bead the inner edge, left-handers the outer.

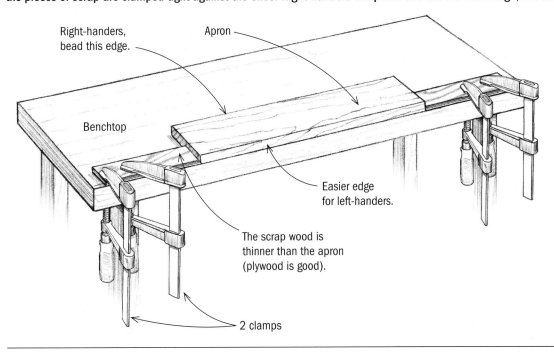

Right-handers, bead this edge.

Apron

Benchtop

Easier edge for left-handers.

The scrap wood is thinner than the apron (plywood is good).

2 clamps

> **WORK SMART**
>
> **C**learly marking the locations of the pocket holes makes it easier to drill them all at once. That way, you can concentrate on setting up the jig and drilling the holes without having to figure out where each hole goes.

location of each pocket hole so you'll know where to drill.

2. Set up your pocket-hole jig as described in "Skill Builder: The Pocket-Hole Screwed Joint" on p. 182, drilling a test hole in a piece of left-over apron stock to make sure the stop collar depth is correct.

3. Drill two pocket holes on the inside of each end of each apron. Use a double-barreled jig to locate the holes or place them about ½" in from each edge, as shown in photo D. The exact location isn't critical. Make sure you don't drill

through the beaded side; the jig goes against the back of the aprons so the pocket holes will show on the underside.

Skill Builder: The Pocket-Hole Screwed Joint

What You'll Need

- Cordless drill/driver
- Pocket-hole jig
- Step drill with stop collar
- 6" #2 square-drive bit
- $1\frac{1}{4}$" pocket-hole screws (washer head #8)
- Scrap wood, $\frac{3}{4}$" thick and some 5/4 stock
- Glue

1. Before setting up the jig or chucking the step drill into your cordless drill/driver, check that the stop collar is in the correct location on the step drill. Insert the bit into the guide bushing on the pocket-hole jig, and position the collar so that the tip is about $\frac{1}{8}$" back from the bottom of the jig, as shown in photos A and B.

2. If you have a bench-mounted jig, simply clamp a piece of $\frac{3}{4}$" scrap in it and drill the pocket holes, as shown in photo D on p. 181. If you're using a portable jig, align the bottom of the jig with the end of the workpiece. Clamp it in place with the supplied clamp (or use a small bar clamp), as shown in photo C. Orient your work so you drill and screw downward.

3. Chuck the step drill bit into your drill/driver and set the clutch and speed for drilling. Put the tip in the

guide bushing and hold the drill lightly so that when you press the trigger, the drill will self-center in the bushing. Drill pushing downward until the collar hits the bushing.

4. Remove the jig and check that there's no hole in the end of the workpiece. If you see any more than a tiny pinprick, set the collar a little lower on the bit to reduce the depth of cut. Try it again and keep adjusting the collar until the step drill makes no hole in the end grain, as the nearest hole shown in photo D.

5. A pocket-hole joint needs two screws (it will pivot around one), so for maximum strength, get them as far apart as possible—about $\frac{1}{2}$" in from each edge. On narrow pieces of wood, they will be quite close together, whereas on a wider piece, they'll be farther apart. A very wide piece could have two sets of two.

6. Switch the drill speed to low, and set the clutch about halfway. Change the step drill bit for a 6"-long #2 square-drive bit. Pocket-hole screws are easier to drive with long bits because of the angle.

7. Make a test by driving a screw into one of the pocket holes without fastening it to another piece of wood, as shown in photo E. Hold this up to the other $\frac{3}{4}$"-thick piece you intend to fasten to it. Make sure the screw will be entirely within the wood and won't poke out the back. If it looks like the screw might be too long, use shorter screws or clamp the jig back slightly from the edge. Remove the screw.

8. Apply glue to the end grain of the piece with the drilled pocket holes, then lay it on the bench with the pocket holes facing upward. Butt the side grain of the other piece against it, as shown in photo F. Drive the screws, adjusting the clutch to get them all the way in. Be careful of overdriving the screw or it may break out the bottom of the drilled piece, as shown in photo G.

You can also use pocket-hole screws to join the edge of one piece to the face of the other, as shown in photo H.

Fasten the Short Aprons to the Legs

1. Clamp a leg at the edge of the bench with the top end near the edge, as shown in photo E. Make sure the inside of the leg is facing up, and use two clamps so it won't shift. Place a piece of ¼" hardboard against the inside edge of the leg and another about 8" farther away from the leg.

2. Apply glue to the end of a short apron, then lay it on the shims, using your thumb to align the top of the apron with the top of the leg.

3. Place a screw in one of the pocket holes, double-check the alignment, hold the apron steady, and drive a 2½" screw into the leg.

4. Drive the second screw.

5. Remove the clamps from the leg and use them to clamp another leg to the bench (don't forget to position it so the leg's best face shows outward in the finished table). Apply the glue, align the apron with the top, and fasten in place, as shown in photo F.

6. Repeat steps 2 through 4 and fasten the other short apron between the remaining legs.

> **WORK SMART**
>
> **S**ometimes the legs are sanded at the factory with the top inch or so slightly narrower than the rest of the leg. If that's the case, there will be a small gap between the apron and the leg at the very top. Don't worry: The overhang from the tabletop obscures any gap.

> **WORK SMART**
>
> **I**t's okay if the apron shifts a little and isn't perfectly aligned with the top of the leg; if it's more than about ⅛" off, though, the table will wobble. Take the joint apart immediately and reposition it. If the screw won't hold, take the joint apart again and put some epoxy or filler in the hole. Reassemble the joint (with glue) after the filler hardens.

Fasten the Long Aprons to the Ends

1. Clamp one end assembly on the right side of the bench so that the apron is vertical. Lay down the ¼" hardboard shims as for the short apron, apply glue, and fasten one long apron to it. Make sure the bead is oriented correctly.

E

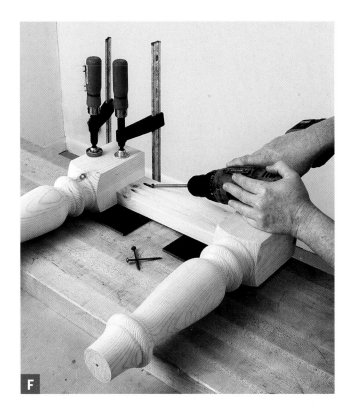

Make and Install the Corner Brackets

1. Select clear leftover pieces of apron stock and cut four brackets, each 8⅜" long. They will span the corner between the two aprons (behind the leg). They provide additional support for the corners, keeping the table square even when it's used hard (see the drawing on p. 175).

2. Cut two pocket holes in each end of each bracket.

3. Roughly mark each bracket on both edges as shown in the drawing below so that you won't have to think about what angle to saw.

4. Set your crosscut saw for a 45° angle, and clamp the block in place. Make the cut so that the block is no shorter (or at worst a very tiny bit shorter). Unclamp, flip the piece end-for-end, and make another 45° cut, as shown in photo H on p. 186.

2. Unclamp the leg and put glue on the end grain of the long apron. Position the remaining end assembly and clamp it in place as before, then position the long apron and fasten it in place.

3. Carefully flip the base and apply glue to both ends of the other long apron. Place the shims, slide the apron in place, and fasten both ends of the apron, as shown in photo G.

Marking the Brackets

Mark the 45° angles on all four ends of the brackets. When you cut the angles with a miter saw, you'll have to flip them end for end. The marks must appear on both sides.

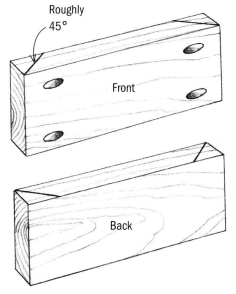

Roughly 45°

Front

Back

Remember: The bracket is longest on the pocket-hole side.

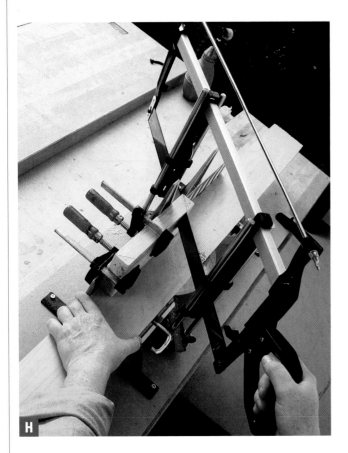

WORK SMART

If the bracket is too short to touch both aprons when you hold it against the leg, use a block plane or chisel to remove the corner of the leg behind the bracket so it can slip into place. It's okay if the bracket is as much as ⅛" away from the leg.

5. Repeat for all four blocks.

6. Hold one bracket in the corner so that the angles touch both aprons and the back touches the leg. Make pencil marks at the top and bottom of the bracket so you can reposition it.

7. Remove the bracket, apply glue to both ends, and put it back in place on the lines.

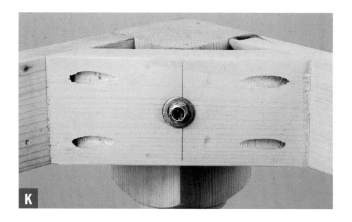

K

WORK SMART

You might need to put a clamp or two on the base to hold it to the bench while you work.

WORK SMART

Driving screws in this order ensures that the bracket is positioned correctly and that it is socked down tight without twisting or breaking it.

ing tape. Change the drill speed to high, set the clutch for drilling, and drill the hole, as shown in photo J.

15. Place the lag screw (with a washer under the head) in the hole and drive with a socket wrench. Drive so it's tight, as shown in photo K, but not so tight the wood crushes around the washer.

8. Holding the bracket, drive a screw in one hole but not fully, as shown in photo I. Driving the screw to full depth at this point might misalign the bracket.

9. Since the next screw goes in the bracket hole diagonally opposite the one you just drove, flip the assembly so the other end of the bracket is down on the bench. It will be much easier to work pushing downward.

10. Check to make sure the bracket is still on its lines, then drive the screw but, again, not to full depth.

11. Drive the other two screws to a similar depth, flipping the piece as necessary.

12. Once all the screws are in place, tighten them in the order in which you placed them.

13. Repeat for the other three brackets.

14. For added strength, drive a 3½" lag screw through the center of the bracket and into the leg. To make the pilot hole, use the pocket-hole step drill. Make the depth about ⅛" less than the length of the lag screw. Set the depth by adjusting the collar or by using a piece of mask-

Level the Aprons and Install Tabletop Fasteners

If all went according to plan, the tops of the aprons are flush with the tops of the legs, but in real life there are always some irregularities.

Planing the tops of the aprons

1. If the aprons are above the legs, plane them flush. If they're below the legs, just leave them. Set the plane for a very light cut; you won't have much to remove. Set the table on the floor and butt it against your bench or a wall, then kneel down to plane. Or if you have something firm to stand upon such as a step stool or box, you can clamp your table to the bench and plane standing up, as shown in photo L on p. 188.

2. To keep the tops of the aprons perfectly flat, use the heel of your plane as a reference. You can use a block plane for this or a larger bench plane if you have one. Rest the heel on the apron near the middle and plane toward one

end, carefully holding the plane level. Skew the plane if required. Don't plane the end grain of the tops.

3. Run the plane over the bracket, and continue along the adjoining apron. Continue in this fashion around the tops of the aprons until they're all flush.

Installing the tabletop fasteners

A wide tabletop shrinks and swells with changes in ambient moisture and must be fastened to take this into account (see Chapter 5 for more information on moisture and wood). Use figure 8–style tabletop fasteners; they rotate slightly to accommodate the movement differential between the tops and the base.

1. Using a Forstner bit, drill ¾" holes in the aprons, as shown in the drawing on the facing page and photo M.

2. Fasten the wide end of the figure 8 in the apron using the screws that came with the clips. Use a 5/64" bit and drill a pilot hole slightly shallower than the depth of the screw to ensure it drives tightly. Drive the screw by hand.

WORK SMART

Practice drilling the figure 8 clip hole in scrap to get the correct location relative to the edge.

Make the Tabletop

The wide tabletop is glued up from narrower boards. With careful gluing and clamping they'll be nearly flat—a little sanding will smooth the joints and add some character.

Selecting the boards

1. Start by taking the time to look at the wood you've selected for the top, and figure out how to orient the pieces so they look best. Choose the most likely boards and crosscut them to about 45", which is the finished length plus 1".

2. Place these boards on the bench and look at the color and grain pattern, paying attention to the color at the edges of the boards. Try to get the colors at the edges to match, as shown in photo N on p. 190, or at least orient them so there are no abrupt changes in color across a glueline, as shown in photo O on p. 190. Also look at knots or defects and decide whether to place them artfully in the top or hide them on the underside.

3. Once the boards are laid out in a pleasing arrangement, draw a triangle across the joints so you don't have to figure it out all over when gluing and clamping (see the drawing on p. 191).

Locating the Figure 8 Clips

Drill ¾" holes just slightly deeper than the thickness of the clip. They should be right on the edge of the apron.

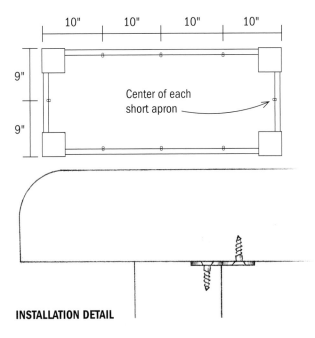

10"   10"   10"   10"

9"

9"

Center of each short apron

INSTALLATION DETAIL

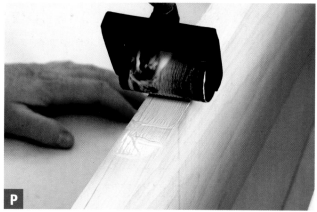

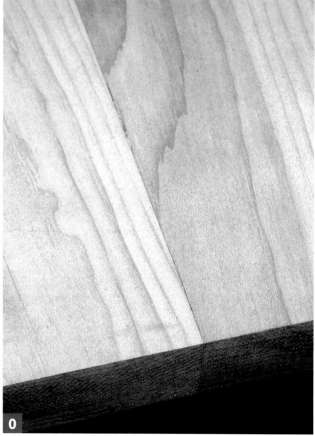

Gluing up the panel

1. Before gluing up the panel, make a dry run to practice clamping according to the techniques shown in "Skill Builder: Clamping Up a Wide Panel" on p. 192. Lay out your clamps and see how this particular set of boards is going to act.

2. Disassemble the dry run and apply glue to both edges of each joint by using a glue roller. Be sure to apply enough glue so that both edges look wet and coated, as shown in photo P. When the panel is clamped up, the glue will squeeze out of the joint in even beads all along its length, as shown in photo Q.

3. Leave the panel in the clamps for at least an hour (at 70°F). Keep it clamped longer if it's cool or damp.

A Triangle Keeps Glue-Ups in Order

Once you've arranged your boards for the nicest-looking panel, draw a triangle that crosses all the joints.

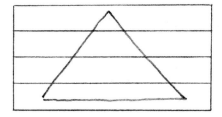

It's easy to see if a board gets out of sequence.

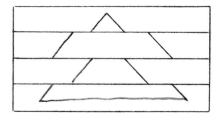

WORK SMART

Don't try to clean up the glue; wiping it will only spread it thinly over the surface of the wood. If you stain your piece, the glue smears will show because the glue prevents the stain from soaking in. Leave the glue to dry and remove it with a scraper later.

R

Cutting the top to length

1. Using a 12" combination square, square from one edge across one end of the tabletop. Make your mark about ⅛" inside the end of the shortest board, as shown in photo R.

2. Place your saw guide along this line, clamp at the bottom, and square up the top edge, as shown in photo S (see "Skill Builder: Making an Accurate Circular Saw Guide" on p. 195).

3. Clamp the top edge of the guide in place.

4. Run the saw along the guide and cut the edge square.

Continued on p. 194

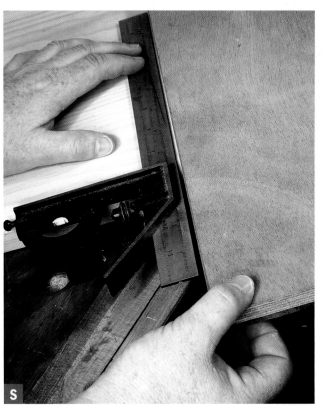

S

Skill Builder: Clamping Up a Wide Panel

What You'll Need

- Five panel clamps that open to at least 30"
- Two clamping pads about 1" × 2" × 48" (or as long as your panel)
- Six or more 12" bar clamps
- Four stout pieces of scrap about 24" long (2×2s are good for this)

Clamping a wide panel is a complex dance between force and geometry. Too much pressure in the wrong place can distort the panel, as shown in photo D, but by the same token you can sometimes force recalcitrant pieces into place with the right clamps in the right locations. That's why it's always a good idea to make a dry run of any clamping setup before the glue goes on. You'll have time to find any problems and work out solutions without the dangling sword of curing glue over your head.

1. Lay out three panel clamps on the bench—one in the middle and one about 2" in from each end. Open the sliding jaw wider than the panel and unwind the handscrew so it's back as far as it can go. If your clamps aren't new, use a scraper to clean up any old glue from the bars so it won't interfere with the leveling or dent your tabletop.

2. Place a long clamping pad against the fixed jaw, and lay the boards on top of the bar. Bring up the sliding jaw with the other clamping pad between it and the wood, as shown in photo A.

3. Press down on the center of the panel, and take up on the middle clamp until it's only moderately tight, as shown in photo B. Next, move to one of the end clamps. Push downward, concentrating your force on the places where the boards don't align. Tighten the end clamp lightly.

4. Clamp across the end of the panel with two stout pieces of scrap and three bar clamps to force the ends

A

B

C

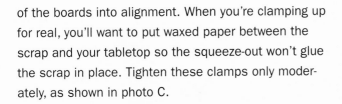

D

7. Check the alignment of the boards in the panel. The joints should be flush or very nearly so. Use a straightedge to check for bowing across the width. If the panel is bowed, ease up on the bottom three clamps, push the panel downward, and take up on the top two. You'll have to try various combinations to see what works. Lift up the assembly and check the underside, too.

8. When the panel is as flat as you can get it, crank all of the clamps as tightly as you can by hand. Start with the middle clamp and then move one to the right. Then go one to the left of the middle, then on to the next clamp on the right. Always work from the center to the ends, alternating sides.

9. Note how the clamps were set (writing it down if need be) and take everything apart to apply the glue. Then reclamp.

of the boards into alignment. When you're clamping up for real, you'll want to put waxed paper between the scrap and your tabletop so the squeeze-out won't glue the scrap in place. Tighten these clamps only moderately, as shown in photo C.

5. Follow steps 3 and 4 for the other end.

6. Next, place two panel clamps with their bars across the top of the panel, and take them up to the same degree as the others.

5. Hook the end of your tape over the freshly cut edge and measure 44" along each edge. Align the guide with these marks, clamp, and saw. If the arrangement of the boards doesn't allow for a full 44", simply cut the top as large as possible. If the overhang ends up a little less than intended, no one will notice.

Rounding the edges

1. Use a block plane or a bench plane to round over the edges according to the method described in "Skill Builder: Rounding Edges with a Plane" on p. 198.

2. Start with the ends of the table. When planing end grain, the far corners sometimes blow out. By planing the ends first, you can fix any blowout in the next step, as shown in photo T.

WORK SMART

Make sure the glue is fully cured before flattening the top. Wet glue expands the wood around the joints, and if you flatten too soon, the joint won't be smooth when it's dry. Wait at least 8 hours to get the best result.

3. Once the ends are done, plane the sides to match, paying careful attention to getting the corners even.

4. When all your planing is done, switch to sandpaper. Don't use a block, cup your hand over the edge so the sandpaper smooths away the plane marks and completes the rounding. Pay close attention to the transitions at the corners, both on the edge and at the top. Shape them by eye to a pleasing contour, as shown in photo U.

Sanding the top

You'll likely find that despite all your care in gluing up, the top is not perfectly flat. Pine tends to move around a lot with changes in moisture levels; it may not be possible to get flat. No matter, you can make it flat enough for this piece by using a random-orbit sander, as shown in photo V on p. 196.

1. Starting with 80-grit paper, move the sander constantly over the surface. Go back and forth, diagonally, in circles; just keep it moving. Pay attention to the high spots, but don't let the sander sit in one spot or you'll quickly make a hollow. Get the top as flat as you can, and make all of the joints smooth, but don't make it perfect or the table will be out of character. Some high and low spots are good for this piece, but they shouldn't be extreme. This step will take 10 minutes to 20 minutes depending on how uneven the joints were.

Skill Builder: Making an Accurate Circular Saw Guide

What You'll Need

- ■ ¼" hardboard about 12" wide and as long as you want the guide to be
- ■ ¾" plywood or MDF about 6" wide and as long as the hardboard
- ■ ¾" screws
- ■ Countersink drill bit
- ■ Cordless drill/driver

This guide makes it easy to use a circular saw to make accurate rips and crosscuts. You won't have to measure any offsets; the edge of the guide is the cut line. Just put a mark on the workpiece at both ends of your cut line. Clamp the guide so the ¼"-thick edge is on the line and the saw and guide rest on the part you want to keep. When you run the saw with its right edge along the MDF, it'll cut right on the line, as shown in the photo below.

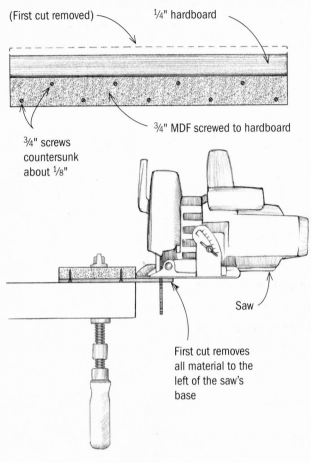

(First cut removed)   ¼" hardboard

¾" screws countersunk about ⅛"

¾" MDF screwed to hardboard

Saw

First cut removes all material to the left of the saw's base

This is such a versatile tool you'll want to make several sizes. I have 24", 38", and 50" guides, plus one that's 8'4" long for cutting sheets of plywood lengthwise.

These dimensions work for my saw; check to make sure they work with yours.

1. Countersink and screw the ¾" MDF to the hardboard, aligning the edges.

2. Clamp the guide to a bench with the hardboard overhanging the edge.

3. Place the right edge of the saw base against the MDF and turn on the saw. Run the saw the length of the guide, being careful to keep the base pushed hard against the MDF. The newly cut edge is the cut line.

V

W

X

2. To remove the random pattern of swirls the sander put in the surface of the wood, replace them with lighter ones by using finer-grit sandpaper. If you don't do this, the swirls will become even more visible, as shown in photo W, when you apply stain or finish. Switch to 120-grit sandpaper, and sand the surface thoroughly to get rid of the 80-grit swirls. Be sure to brush the surface to clean away the coarser grit before sanding with finer paper.

3. Use 180-grit paper and sand again, then sand with 220 grit. These steps will take only 3 minutes to 5 minutes each.

4. Finally, finish by sanding back and forth along the grain by hand. Use a block and 220-grit paper until there are no swirl marks.

5. Brush off the bench, flip the top, and smooth the underside of the top using a random-orbit sander. Start at 120 grit and work up to 220 grit, following the sequence above.

6. Break the sharp edge on the underside of the tabletop by sanding lightly using a hand-sanding block at an angle with 220-grit paper, as shown in photo X.

> **WORK SMART**
>
> **A**lways plane before sanding. Once you've sanded, don't use a plane—little bits of grit left by the sandpaper will quickly dull your plane iron.

Sand the Base

Now is the time to sand the rest of the table and clean it prior to putting on the finish. Once the top goes on, it'll be a lot harder to get at the inside and underside of the table.

1. Sand the aprons and legs by hand, smoothing the surfaces and breaking all of the edges. Start with 120 grit on the aprons and 220 grit on the legs. If the surface of the aprons is particularly rough or dirty, you can use a random-orbit sander, but don't let the edges of the disk hit the legs—they'll cut a groove.

2. Take some time working on the inside of the base, breaking all of the corners and smoothing the surfaces.

3. Pay attention to the corners of the legs, rounding them slightly. Be careful not to let the sandpaper scratch the aprons while doing this.

Cleaning

1. Before applying any finish, vacuum the table with a brush attachment and clean up your work area, too. Dust left on the bench, floor, and surrounding surfaces will swirl into your finish as you move around.

2. Wipe the entire coffee table with a loosely wadded tack rag.

Finishing

No matter what finish you plan to put on pine, for a fine finish, the first step after sanding is to "condition" the surface with shellac. Since pine is a difficult wood to finish well, a light sealer coat of shellac solves problems such as blotchy stain, knots bleeding through paint, dull natural color, and resin beading out. In photo Y, the board on the left has been shellacked; on the right, the same stain was applied to the bare wood.

Prestain conditioning

1. Thin dewaxed shellac with alcohol about 50%. If you're making your own, make up a mixture that's proportional to ½# of shellac to 1 gal. of alcohol (see pp. 140–141 for more on shellac).

2. Using a brush, apply a thin coat to the top, as shown in photo Z. I prefer to leave the legs and the aprons unshellacked to save time and trouble. The end grain in the legs will stain a deeper color, giving more character, and the blotchiness won't really show on the aprons. Let dry at least ½ hour.

Continued on p. 199

What You'll Need

- Sharp block plane or bench plane
- 5/4 pine at least 12" long
- Vise
- Sandpaper

E ven though a block plane is used mostly to make things flat, smooth, and square, you can also use it to make a roundover. It's only a matter of planing a series of flats and then breaking the corners with sandpaper, as shown in the drawing on the facing page.

1. Plane a chamfer at 45°. The first few cuts will produce a very narrow shaving that gets wider with each cut.

2. Plane until the chamfer is about ½" wide, as shown in photo A. You can check the evenness of your planing by making sure the chamfer is ½" all along its length.

3. Move down to the lower edge of the chamfer and hold the plane at about 45° to it. Plane away the shoulder of the chamfer until it's about ⅛" wide. Then move to the upper corner of the chamfer and do the same, but make the new chamfer narrower, as shown in photo B.

4. Next, plane the shoulders made in step 3, then plane the shoulders that result from that cut. Finally, you'll be planing very narrow chamfers and the edge will be nearly round.

5. Finish rounding the edge using a piece of sandpaper. This time, don't use a hand-sanding block—cup the sandpaper around the edge to get it round.

1. Plane away the corner.

2. Plane away the lower shoulder.

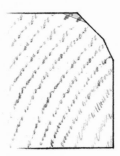

3. Plane away the upper shoulder.

4. Plane away the second shoulders.

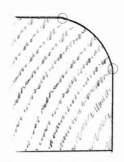

5. Plane away the knuckles.

6. Hand-shape to final shape.

AA

3. After applying the sealer coat, the surface seems rougher than when you began because the finish raises the grain, so sand very lightly using 320-grit paper and a sanding block. Remove the nibs without sanding through to the bare wood, and vacuum up the dust when you're done.

WORK SMART

Always test your stain on an offcut from your project that's been sealed and prepared in exactly the same way as the project. Color charts are approximate. The only way you'll know what the finish will look like is to test it on the same wood, let it dry properly, and look at it in the sunlight.

Applying the stain

Using a water-based stain and following the manufacturer's directions, stain the entire table, inside and out, top and bottom. Work only a small area at a time, and keep plenty of paper towels on hand to wipe up.

Applying the wax

To make new furniture look like it's had good care for generations, use a dark-colored wax over the stain.

1. Apply the wax with a square of a white synthetic steel-wool pad, rubbing in circles.

2. Wipe down using cotton rags or lint-free paper towels, as shown in photo AA on p. 199.

3. Apply two coats to the base and the underside of the top. Apply three coats to the top.

Affix the Top

1. Put a piece of clean carpet, a blanket, or a towel down on the bench, and place the tabletop on it upside down.

2. Set your sliding square to 2" and make light marks in from both the ends and the edges at each corner, as shown in photo BB.

3. Place the base on the tabletop so the corners of the legs align with these marks.

4. Make light marks to record this alignment, in case the table gets bumped while you're working.

5. Drill a $5/64$" pilot hole, and drive one of the screws that came with the figure 8 clips into the tabletop. Use a hand driver, as shown in photo CC.

BB

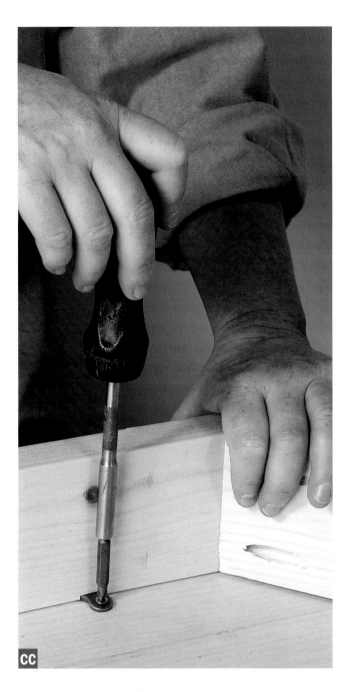

6. Check the opposing corner to make sure the base is still aligned to the marks. If it isn't, rotate the base a little to get it in place. When aligned, drive a screw into a nearby clip.

7. With the base locked in position, drive the rest of the screws.

8. Apply one more coat of wax to the tabletop before using it.

9. Maintain your coffee table by keeping it clean and applying a new coat of wax once or twice a year.

WORK SMART

In many cases, the base aligns with three corners perfectly, but the fourth is off a bit (this is likely to happen if your tabletop ended up a little shy of 44" long). Shift the base to distribute the difference between two or more corners. Instead of one corner being way off, make them all a little bit off.

Classic Bookcase

"A room without books is a body without a soul," wrote Cicero more than 2,000 years ago. A great scholar, he also understood the pleasures of building bookshelves, although a noble statesman of ancient Rome couldn't be expected to do handwork. He wrote to a friend, "Your men have made my library gay with their carpentry work . . . Now that [we have] arranged my books, a new spirit has been infused into my house . . . Nothing could look neater than those shelves."

Things haven't changed much—even in the Information Age you still can't have too many bookcases, and this classic design will instill new spirit into any room. Its simple good looks work just about anywhere. With a neat finished back, it can serve as a simple room divider, or it can go against the sofa or wall. Paint it white, use bright primary colors for the kids, distress it, or use crackle paint to make it look old, or dress it up with fancy moldings.

This bookcase is designed to get you up to speed with one of woodworking's most valuable tools—the router. While building this project, you'll learn two fundamental ways to guide a router, how to make a nearly foolproof jig that keeps the router in check, and how to use four different router bits. You'll learn how to make grooves across the middle of a board (called dadoes), make a shoulder in the edge of a board (a rabbet), rout edge profiles, and use straightedges and templates to put a smooth edge on a piece of wood—even one with curved edges.

What You'll Learn

- Setting up and using a router
- Routing rabbets
- Template routing
- Making dadoes with a router
- Squaring the ends of a routed dado or groove
- Routing edge treatments
- The elements of mitering a molding
- Filling and smoothing using putty
- Prepping for and applying a high-quality paint job

Everyone's heard Murphy's Law—if anything can go wrong, it will. It's not a bad motto for woodworkers—if you interpret it the right way. Though often cited as a curmudgeonly joke, the real meaning of Murphy's Law is that with experience and forethought you can get the outcome you want.

Murphy's Law was developed in 1949 by scientists and engineers of the U.S. Air Force researching the effects of rapid deceleration on the human body. The tests were technically complex and brutal on the subjects. Every time a technician plugged an electrode into the wrong receptacle or made an ambiguous note, the whole operation had to be repeated at great expense and a fair bit of human suffering. The team (led by Captain Edward A. Murphy) soon developed a methodology that required analyzing every aspect of the test to find everything that could possibly go wrong. They then redesigned the whole system so nothing could go wrong and the tests could be conducted quickly and safely.

Knowing how to cut and fit mitered moldings is useful for all kinds of woodworking, from furniture making to picture framing and from putting up moldings to trimming windows.

Classic Bookcase

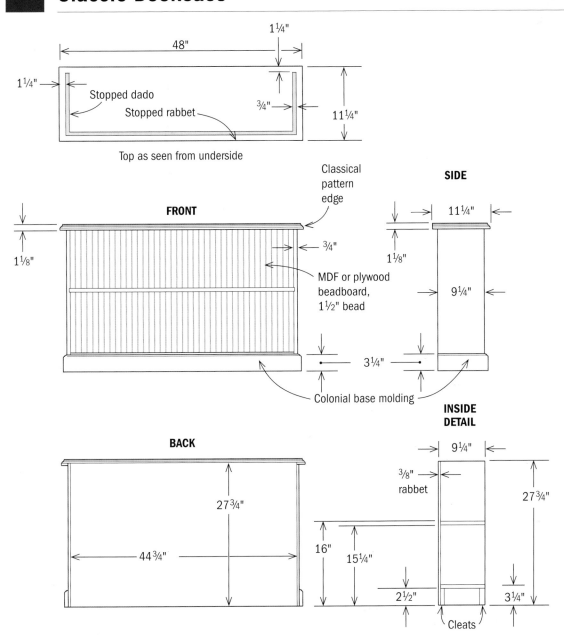

48"

1¼"

1¼"

Stopped dado

Stopped rabbet

¾"

11¼"

Top as seen from underside

Classical pattern edge

FRONT

1⅛"

¾"

MDF or plywood beadboard, 1½" bead

SIDE

11¼"

1⅛"

9¼"

3¼"

Colonial base molding

INSIDE DETAIL

BACK

27¾"

44¾"

9¼"

⅜" rabbet

27¾"

16"

15¼"

2½"

3¼"

Cleats

Variation: Information Age Classic Bookcase

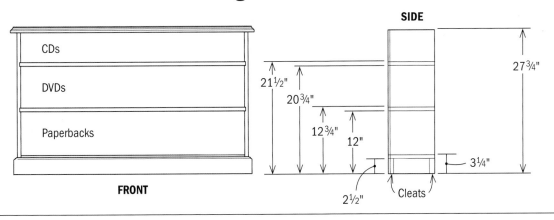

CDs

DVDs

Paperbacks

FRONT

SIDE

21½"

20¾"

12¾"

12"

2½"

Cleats

27¾"

3¼"

MATERIALS

Quantity	Description	Actual Dimension	Length	Notes
1	Top	1⅛" × 11¼"	4'	5/4 × 12" poplar Finished at 48"
1	Sides	¾" × 9¼"	6'	1 × 10 poplar Yields two sides, finished at 28"
2	Shelves	¾" × 9¼"	8'	1 × 10 poplar Yields two shelves, finished at about 44"
1	Colonial base molding	½" × 3¼"	8'	Any material; primed is nice.
1	Aprons and cleats	¾" × 2½"	10'	1 × 3 poplar Two pieces finish to 44"
1	Back	¼" × 4' × 4' (half sheet)		MDF (primed is easiest to paint) or plywood beadboard, 1½" beads, or plain MDF or plywood
1	Molding stock	¾" × ¾"	18"	Used for converting jig for stopped moldings
	Yellow glue			
	#7 or #8 flat-head wood screws		1¼"	Self-drilling, square drive preferred
	#7 or #8 flat-head wood screws		2"	Self-drilling, square drive preferred
	Brads		1"	
	Double-sided tape			
	Quick-dry, easy-to-sand filler			
	Paintable caulk			
	Random-orbit sanding disks			150 and 220 grit
	Sandpaper			120, 180, 220 grit
1	Tack cloth			For use with oil-based paint only
1 quart	Primer			Optional, gives the best finish
1 quart	Paint			Oil or water based
	Solvents			As per paint label
	Denatured alcohol			For general cleanup
	Paintbrushes			Five disposables or one moderate-quality bristle brush if using water-based paint

TOOLS

- Crosscut saw capable of cutting an 11¼"-wide board
- Router
- ½"-dia. × 1" cutting depth template bit
- ½"-dia. × ¾" cutting depth template bit
- ⅜" rabbet width × ½" cutting depth rabbeting bit
- Router depth gauge or small sliding square
- Tape measure
- Several small bar clamps, two 24" panel clamps, and two 48" panel clamps
- Plywood straightedge
- Steel ruler
- 6" sliding square
- 12" sliding square
- Classical pattern edge-trimming bit
- Corner chisel
- Mallet
- #8 countersink drill bit
- Square-drive bit
- Cordless drill/driver
- Warrington hammer
- Circular saw and 50" guide
- Nail set
- Random-orbit sander
- Hand-sanding block

You'll need four router bits to build the bookcase. From left: a ¼"-shank template-cutting bit with a 1" cutting depth and a ½" dia.; a classical pattern bit with a ½" shank; a ¼"-shank template-cutting bit with a ¾" cutting depth and a ½" dia.; and a ½"-shank rabbet-cutting bit with a ⅜"-wide cut.

Successful woodworking is all about adopting the true spirit of Murphy's Law—that you can plan ahead and prevent problems before they occur. With a little experience gained from time in the shop and from reading and talking with other woodworkers, you'll soon learn the kinds of things that can go wrong and how to prevent them. You'll get lots of practice with this when you're using a router.

Routers are one of the most versatile tools in the shop, but I suspect more woodworking projects have been messed up by routers than any other means. It's not that routers are particularly dangerous or malignant; they are so fast and efficient that a moment's inattention can mess up hours of previous work.

This bookcase gives you a good grounding on using a router with complete control. You'll learn to guide the router with fences, bearing-mounted bits, and jigs so that the machine cuts along the path you want.

You'll use the router to make dadoes and rabbets to build a bookcase that uses standard dimensional lumber, goes together square, is nearly impossible to rack, and—with its complex profile routed on the top edge—is classically good looking.

You'll also learn the basic elements of a high-quality paint job, mitering corners, and applying molding—all skills you can use around the house, from refinishing furniture to building picture frames or trimming windows.

Building the Bookcase

Take the time when selecting your materials for this project to get wood that has very little cup. A cupped ¾"-thick board simply will not slide into a ¾" dado.

WORK SMART

If your miter saw or chopsaw can't handle boards of this width, use a circular saw and crosscut guide.

Cut the Parts to Length

1. Check your material for knots, gouges, and other flaws. Orient flaws on the inside or back edge whenever possible.

2. Crosscut two ¾" × 9¼" sides to 27¾". Cut the 1⅛" × 11¼" piece for the top to 48".

3. For ease of handling, rough-cut the ¾" × 9¼" shelves to 48".

Lay Out the Dadoes and Rabbets

To avoid confusion later and to understand the bookcase-building process, start by drawing directly on the wood all the various cuts you'll make with the router. This way you'll see clearly on the wood just what you're supposed to do, and it's apparent immediately if the router is set up wrong or if you're about to rout in the wrong place.

Laying out the sides

The shelves slip into dadoes cut into the sides. For the shelves to be parallel to the floor and to one another, the dadoes in each side piece must be exactly the same distance from the floor. To make sure this happens, you'll mark and cut them together, with the bottom edges of the side pieces aligned. You'll also mark the locations of the cleats that will be fastened beneath the lower dado to give extra support to the bottom shelf and provide a firmer footing for the bookcase.

Each side piece also has a rabbet on the inside back edge, running the full height to conceal the edges of the MDF back. Building it this way gives the bookcase a more polished appearance and provides solid bearing surfaces for the back, preventing racking.

1. Lay the side pieces on the bench side by side with the ends flush. Clamp them together and down to the bench so they can't shift.

2. Hook the tape measure from the bottom, and lay out the locations and widths of the cleats and the dadoes according to the measurements in the drawing on p. 205. Put hatch marks where the dado will be cut to make it clear where material will be removed, as shown in photo A.

3. Unclamp the side pieces and draw a line ⅜" in from the back edge of each piece. The side pieces are mirror images of one another, so the back edges are the edges that are touching in the middle. Put hatch marks in this area to represent the rabbet.

A

Laying out the top

The top of the bookcase is wider than the side pieces, and it overhangs all around—1¼" on the sides and front and ¾" at the back. The side pieces slip into stopped dadoes in the top so that the groove won't show at the front of the case. The back edge of the top is also rabbeted to accept the MDF back, but unlike the sides, it has a stopped rabbet. A rabbet that went the full length of the back would mar the bookcase's appearance from the side.

1. Use a sliding square to set out the stopped dadoes according to the measurements in the drawing on p. 205. Draw hatch marks in what will be the dado.

2. Draw the stopped rabbet along the back edge, with hatch marks to show the material that will be removed.

Rout the Rabbets

Now that you know exactly where the rabbet is cut, you can use your router and rabbeting bit to remove some wood.

Cutting rabbets for the sides

Set up your router as described in "Skill Builder: Router 101—Rabbeting" on p. 212, and cut ⅜"-wide × ⅜"-deep rabbets in the back edges of the side pieces.

Cutting rabbets for the top

The stopped rabbet in the top is ¾" wide × ⅜" deep—wider than the rabbets in the sides. This is to allow for the fact that the top overhangs in the back. You'll need two setups to cut it.

1. Clamp the top to the workbench with the hatch marks up. The edge to be rabbeted should overhang the bench. Make sure the clamps are out of the router's way.

2. Don't try to start the rabbet perfectly at the left corner of the stopped dado on the left side of the top. Simply start the rabbet within the hatch marks of the stopped dado—the left edge will be cut cleanly in the course of a later step when you route the stopped dado. Continue routing until you reach the other stopped dado. Stop when the rabbeting cutter is within the hatched area that will be removed when routing the other stopped dado.

3. Widen the rabbet to ¾" by making a second cut with a ¾"-long template-routing bit according to the method described in "Skill Builder:

WORK SMART

Both ends of these rabbets will be hidden from view, so don't worry if they aren't perfect.

B

WORK SMART

Keep the router base on the workpiece until the bit stops turning.

Template Routing with a Straightedge" on p. 216, as shown in photo B. Set the depth to ⅜". Once more, the stopped dadoes will clean up the ends of the rabbet.

Rout the Dadoes

Dadoes are a great way to support bookshelves. They're good looking and strong, and when the fit is good, the square corners lock the shelves in place to prevent racking. There are many ways to cut dadoes, but using a router jig that takes Murphy's Law into account makes the

operation safe and quick. Since the jig uses a template-routing bit and has guides on both sides of the cut, the bit is under control at all times (see "Skill Builder: Making a Dado-Cutting Jig" on p. 218 for how to make this jig).

Cutting dadoes for the sides

1. Clamp the two side pieces together with the penciled dadoes aligned and the rabbets in the middle. Clamp them to the bench with the penciled dado about 8" from the end and one of the rabbets slightly overhanging the front edge.

2. Place the jig on the workpiece so the left edge of the slot aligns with the left side of the dado. The hatch marks should be visible in the slot. The right side of the slot should be parallel to the right side of the penciled dado but slightly wider. If the slot isn't parallel, either the jig's vertical fence isn't square to the slot or the dado was drawn wrong. Figure out which is the case and fix it.

3. Make sure the vertical fence is snug against the edge of the workpiece by using long clamps tightened across the two pieces, parallel to the slot, as shown in photo C. Clamp the tee at the top end of the jig down to the bench.

C

D

Cutting dadoes for the top

Dadoes don't have to go the full width of the piece. You can use this jig to stop the dado at any point by slipping a ¾" × ¾" × 18" piece of scrap into the slot. Here's how it works: As you're routing along the left edge of the slot, the bearing encounters the stop and prevents any further forward motion. Slide the bearing against the right side of the slot and bring the router home. This leaves a stopped dado with rounded corners. Later you'll chisel them square.

Rather than measuring to find the proper length for the slot, follow this procedure.

1. Clamp the jig in place square against the edge of the workpiece and over the penciled dado. Because the jig is near the edge of the workpiece, you can't get clamps on both sides of the jig, so clamp across only one side and clamp the end to the bench, as shown in photo D.

2. Using a sliding square, make a vertical mark on the inside of the slot and a horizontal one on the face of the jig at the end of the dado, as shown in photo E on p. 215.

3. Put a piece of ¾" × ¾" molding stock that's clearly too long into the slot, and push it up against the top end (it'll be a little loose in the slot but that's okay). Draw a line on the molding that aligns with the mark you made in step 2. Crosscut the molding to length, and check that its end aligns with the mark on the jig.

Continued on p. 215

4. Set the depth of cut to about 3⁄16", guide the router into the slot, and rout across both side pieces, running the bearing against the left side of the slot. Once you've routed both pieces and the bit has traveled beyond the far edge of the workpiece, slide the router to the other side of the slot so the bearing runs on the right side of the slot and pull it toward you.

5. Stop the router, set the depth of the final cut to 3⁄8", and make another run around the slot.

6. Repeat this procedure on the other dadoes.

What You'll Need

- Router
- Rabbeting bit, $\frac{3}{8}$" rabbet depth and a $\frac{1}{2}$" cutting length
- Scrap wood for practice, 1" × 5" × 3'
- Two clamps
- Combination square or router-bit depth gauge
- Dust mask
- Hearing protection
- Safety glasses

Rabbeting is a great application to start learning to use the router because it's a simple cut that's easy to set up and guide. With the bearing on the tip of the rabbeting bit, the router is docile and easy to control as long as you remember this rule: Rout against the turning of the bit. The easiest way to figure out just which way to rout is by making an L with your right thumb and forefinger. Point to the surface you want to rout with your thumb. Your right forefinger points in the direction the machine should travel.

If you rout in the other direction (called climb cutting), the router tends to pull itself along by the turning of the bit, making it difficult to control. Climb cutting is used in some circumstances, but most of your routing should be done using the above rule.

The key to router success is keeping the base flat on the surface. If the router tips just a little off vertical, it can rout a perfectly molded divot in the edge before

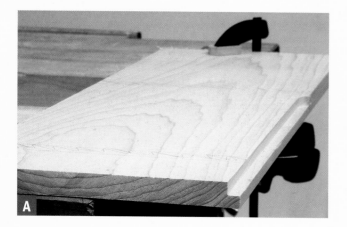

A

Routing Rule of Thumb and Forefinger

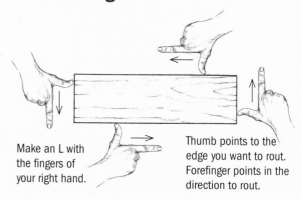

Make an L with the fingers of your right hand.

Thumb points to the edge you want to rout. Forefinger points in the direction to rout.

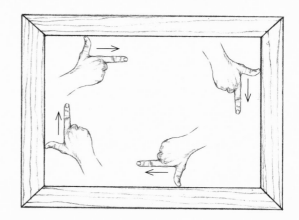

you know what has happened. Concentrate on keeping the router flat on the work, with the bearing pressed against its guiding surface. Keep your work area clean, your clamps out of the way, and make sure the router cord runs free before you start. That moment of inattention when you look down to step over an obstruction could be enough to mess up a perfect edge. Whenever you run a router, you should wear hearing protection, eye protection, and a dust mask.

Setting Up the Router

Before working with the collet or bit, be certain the router is unplugged.

1. Move the base plate away from the body of the router to give you room to work. Loosen the collet locknut by holding the shaft steady with one of the

B

two wrenches that came with your router, as shown in photo B, or by using the shaft-lock button if your router has one.

2. Insert the shank into the collet as far as it can go, then back it out about $\frac{1}{16}$" and hand-tighten the collet.

3. Using the wrench or wrenches, crank the locknut as tightly as you can.

4. Put the router upside down on the bench. Hold the $\frac{3}{8}$" step on a router-bit depth gauge over the side of the bit and raise or lower the cutter until the end of the cutter just touches the gauge, as shown in photo C. You can also use a sliding square. Set the blade at $\frac{3}{8}$" and put the end of the blade on the router base, with the base of the square as a height gauge for the bit. Make sure the end of the blade is within the square's base so that it can rest flat on the router base plate so the base of the square is perpendicular.

5. Check that your workspace is clear, make sure the on/off switch is set to off, and plug in the router.

Continued on p. 214

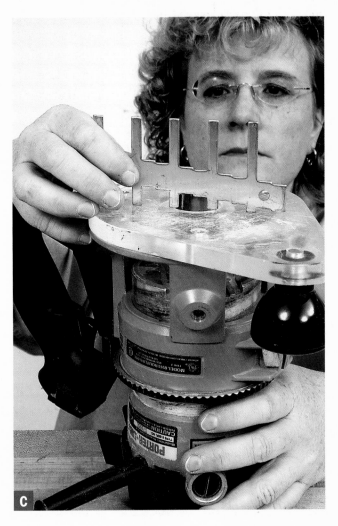

C

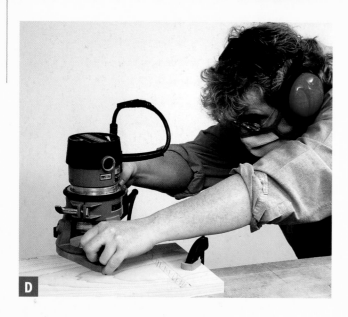

D

Routing the Rabbet

1. Clamp the workpiece with the edge to be rabbeted overhanging the edge of the bench (this prevents the bearing from scoring your benchtop). Position the clamps so the router base won't run into them when making the rabbet.

2. Start your first test cut anywhere in the middle of the piece. Put the router base on the workpiece, but make sure the bit is about 1" away from the edge. The seemingly great distance from the work is because many routers give a little twitch when switched on, and you don't want the bit to accidentally touch the wood before you're ready to cut.

3. When you're routing, you need to be in a strong and stable position that lets you see what's going on at the bit. Stand well back from the work and bend at the knees to see the cutter, as shown in photo D.

4. Once the router comes up to speed, push it directly inward. It'll make a lot of noise and dust at first, but as soon as the bearing touches the edge, it'll quiet down. Push the router from left to right, at the same time exerting a steady inward pressure to keep the bearing against the edge. Don't push too hard, or the bearing will dent the wood.

5. Make a cut several inches long, move the bit about 1" away from the edge (as in your starting position), and turn off the router. When the bit has stopped spinning, remove the router and check the depth of cut with a sliding square. Adjust by trial and error until the depth is correct.

Rabbeting the Ends

It takes some practice to get the rabbet perfect at the ends of the board. The most common mistake is taking a little chunk out of the end by running around the corner. This happens when you are trying too hard to keep the bearing in contact with the edge of the workpiece. With a little practice, you can get perfect corners. The secret is 1" of climb cutting at each end.

1. Clamp a board as described above, and start the router about 2" to the right of the left end of the edge to be rabbeted.

2. Turn on the router, push it against the edge, and slowly bring it to your left. You're making a climb cut in the opposite direction of the rule of thumb and forefinger. You'll find the router doesn't want to stay against the edge as it does when you rout in the other direction. Be prepared for the router to pull toward you a little, but don't worry if it does. Go slowly and you'll be in control. Watch the bit, and you'll see that before the bearing reaches the left end of the board, the wider diameter cuts a rabbet right to the end.

3. When you reach this point, stop your motion to the left, and push the router inward so the bearing contacts the edge. Then you can start cutting from left to right in the normal fashion.

4. When you get to the far end, slow down and watch what's happening. In a similar fashion, you'll stop when the cut goes to the end but before the bearing runs off the edge. The biggest mistake people make when routing is to assume they have to rush just because the router is so fast and noisy.

WORK SMART

Since this bookcase is going to be painted, you could forego the clamping and simply screw the jig to the workpiece. The screw holes won't show once they're filled with putty and sanded smooth. Be sure to use flat-head screws and get the heads below flush so they won't trip up the router.

4. Cut the stopped dado in two passes, just as you did the dadoes in the side pieces. Start the router as you did for the sides, and run the bearing down the left edge of the slot. When it reaches the molding stop, push into the stop and over to the right side of the slot. Then pull the router toward you, keeping the bearing against the right side of the slot.

Rout the Edge Treatment on the Top

To give the top edge of the bookcase more visual interest, rout a profile in the front and sides using an edge-trimming bit with a tip-mounting bearing. The process is the same as

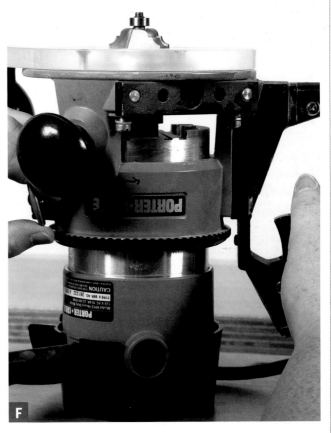

that used to make the rabbets, including starting and stopping without turning the corners.

1. Set the depth of cut by testing it on offcuts that are the same thickness as the top of the bookcase. Clamp the offcut to the bench and put a classical pattern bit into the router. Flip the router over and set the depth by eye so the corner of the bit is about 1/16" above the base, as shown in photo F.

2. Run this bit setup on the test piece for a few inches to see if you like the way the profile looks. Experiment with raising and lowering the bit slightly for different effects.

WORK SMART

Position a clamp so the router base bumps into it before the bearing plows a groove in the side of your bench.

SKILL BUILDER: Template Routing with a Straightedge

What You'll Need

- Router
- Template router bit with a ¼" shank, a ½" diameter, and a ¾" cutting length
- Plywood or MDF fence, ¾" (or ½") × about 6" wide × 50" long
- ¾" × ¾" × 18" molding stock for stop
- Sliding square
- Four clamps
- Safety gear

Straight-sided router bits with a bearing mounted on the shank are called template bits. They're most often used with a template (straight or curved) clamped atop the workpiece. The bearing runs against the template, and the bit cuts the material below it exactly flush with the template.

As always when using a router, be sure to wear a dust mask, hearing protectors, and safety glasses.

Setting Up

Use a piece of plywood or MDF with a perfectly straight and smooth edge that's a little wider than your router's base and at least a couple of inches longer than the edge to be routed. Make sure there are no dings or voids in the edge—the cut edge will exactly match the template's edge, including flaws.

1. Clamp the workpiece to the bench and use a sliding square to mark ¼" in from the edge on both ends of the workpiece.

2. Clamp your straightedge along this line, with the clamps well back from the edge so they won't interfere with the router.

3. With the bit secured in the collet, place the router on the left end of the fence where it overhangs the workpiece, and raise or lower the bit until the bearing

Setting Template Router Bit Depth

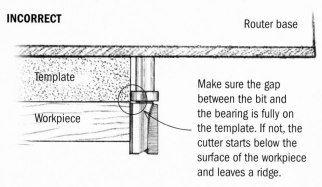

INCORRECT

Router base

Template

Workpiece

Make sure the gap between the bit and the bearing is fully on the template. If not, the cutter starts below the surface of the workpiece and leaves a ridge.

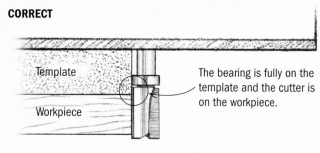

CORRECT

Template

Workpiece

The bearing is fully on the template and the cutter is on the workpiece.

and the space between it and the workpiece are wholly on the fence, as shown in the illustration above, or it leaves a ridge.

The lower end of the bit should extend below the workpiece—but it doesn't matter by how much.

Template Routing

1. Start the router well away from the edge, get your stance, and push the bearing against the template. Rout from left to right, as shown in the photo at right.

2. Listen to the router to judge the feed rate; it will tell you what to do. If you're going too fast, you'll hear the motor slow down a little. If you feed too slowly, the surface of the wood will burn.

3. If your router seems to complain too loudly, don't try to cut the full depth in one pass. Slide the router back 1" or 2" so it's not cutting, and turn it off. Raise the bit a little and make a cut the full length of the

workpiece. Then lower the bit to the final depth and make another full-length pass. Make sure that after you raise the bit, the bearing still rides on the fence.

3. When you like the depth setting, clamp the top to the bench (dadoes down and rabbet toward the back of the bench) with the left end overhanging about 18". Start routing at what will be the left back corner of the top. Rout down the left end, around the corner (go slowly and keep the bearing in contact), and down the front until you get close to the bench, as shown in photo G on p. 219.

4. Reposition and reclamp the top (twice if necessary) to rout all the way around it—right up to the back right corner. When resuming your routing after repositioning, back up a few inches and start the router away from the edge. Push inward, and make the bearing contact the edge at a place you've already cut. Move the router to the right and continue cutting the profile. You'll never see where you started and stopped.

Square the Stopped Dadoes

The rabbet has a rounded front edge, but your side pieces have square corners. Make the stopped dado fit the sides by squaring its ends using a chisel. You can make a square corner with a bench chisel, but it's far easier with a proper corner chisel.

1. Put the arms of the chisel against the end of the dado made by the router, with the corner against your penciled layout marks.

2. Hold the chisel perfectly vertical, as shown in photo H on p. 219, and whack it with a mallet. Clean out the excess down to the bottom of

> **WORK SMART**
>
> **P**ractice with the corner chisel before working on your bookcase.

What You'll Need

- ¾" MDF or plywood 4" to 26" wide
  (two pieces 30" long, two pieces 20" long)

- Plywood or MDF, ¾" × 1" × 4"
  (scrap okay, size is approximate)

- Stop, ¾" × ¾" molding stock, about 18" long

- Two bar clamps

- Two panel clamps (must open at least 30")

- Pocket-hole jig

- 1¼" pocket-hole screws

- 1¼" flat-head screws

- Combination square

- Sticky notes

- Block plane and/or hand-sanding block

- Two offcuts from shelf material, each about 6" long

- Backsaw

Murphy's Law Dado Jig

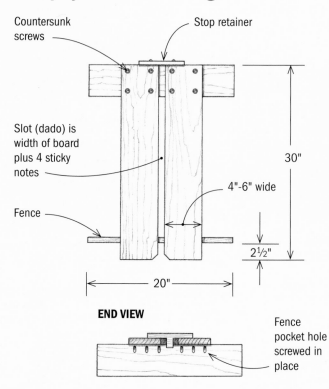

There are several ways to cut dadoes, but the safest, easiest, and most precise way is to use a jig like the one shown here. Clamp the jig down tightly with the fence against the edge of the board, and it makes identical dadoes that are always square to the edge and exactly the right width. You'll make two passes with the router to make each dado—the first with the bit set to cut into the wood to a depth of about ³⁄₁₆", the second to cut to the full ⅜".

It's critical that the fence on this jig be square to the slot. The sequence of assembly is important for achieving this result.

1. Draw a line perpendicular to the edge of one of the 30" pieces. Using a 12" combination square with the base against a long edge, make the line about 2½" back from one end.

2. One of the 20" pieces is the fence that fastens on its edge with pocket-hole screws along the line you just drew. Using the pocket-hole jig, drill pocket holes

in the piece, as shown in the drawing above. Place the fence on the line and drive a pocket-hole screw at just one end. Make sure the other end is also on the line, perfectly square to the edge. Fasten the other end in place. Double-check using your sliding square to make sure the fence is perpendicular to the slot, and drive the remaining pocket-hole screws.

3. Fasten the base (the other 20" piece) face down to a line squared across the other end about 4" back from the edge.

4. Flip the piece over, with the fence against the edge of the bench. Peel two sets of two sticky notes from a pad and stick one set to each end of each shelf offcut. The sticky notes act as shims to make the slot a little wider than the shelf. Position the shelf offcuts along the inside long edge of the 30" piece, with one offcut at each end.

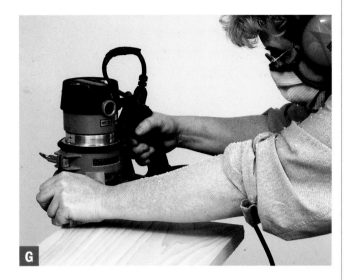

5. Place the other 30" piece alongside the offcuts. Align the ends with the other 30" piece, and clamp the two long pieces together near both ends, as shown in the photo above. This puts the second long piece perfectly parallel to the first, separated by the shelf offcuts. Fasten the fence and the base to the second long piece with the appropriate screws. Don't use glue because you might want to make an adjustment later. Remove the clamps and the offcuts.

6. The jig needs a retainer across the slot at the top end so you can slip a stop in place to limit the length of the dado (see "Cutting dadoes for the top" on p. 211 for details on fitting and using the stop). Use a small piece of scrap (a piece of leftover hardboard is perfect), and fasten it with a couple of pocket-hole screws. Their washer heads are perfect for this, since you won't have to countersink them into the thin hardboard.

7. Using a backsaw, cut away the inside corners of the long pieces, widening the mouth of the slot, which makes a good place to start the router.

8. Break all of the edges using a block plane or hand-sanding block so the jig is easy on the hands and has a neater, more finished appearance.

the dado, using a bench chisel if necessary, as shown in photo I on p. 220.

3. Check the fit of the side pieces in the dadoes. If the fit is tight, it's probably because there's some cup in your side pieces. First try planing

WORK SMART

If the fit of the dado is tight, it's easier to alter the board that fits into the dado rather than the dado.

a chamfer around the upper end of the side piece. If that doesn't work, use a block plane to take a few strokes off one or both faces of the upper end where they slip into the dado.

Fasten cleats to the sides

1. Now work on the lower ends of the sides. Use a sliding square to make a mark ¾" from the front edge of the bookcase and another mark ¾" in from the rabbet on the back edge. Measure the distance between these lines.

2. Crosscut two pieces about 10" long from the ¾" × 2½" poplar, and clamp or tape together and cut two identical pieces to fit between the marks. These are the cleats.

3. Clamp a cleat to one side piece so its top edge is flush with the bottom of the dado, as shown in photo J.

4. Turn the piece over and drive three screws to hold the cleat in place. Drill pilot holes from the inside out. Don't countersink too deeply. The heads need to be only about ⅛" below the surface.

5. Remove the screws and clamps, apply glue to both surfaces, reclamp, and fasten in place. Wipe up any glue drips with a wet paper towel.

Drill Pilot Holes All Around

To increase the strength and stiffness of the bookcase, the dadoes are reinforced with long screws into the end grain. Rather than guessing where the screws go or measuring to find their location, just drill the pilot holes before assembly—from the outside in, just as you did when installing the cleat on the back of the Outdoor Easy Chair on p. 164.

1. Drill three fairly evenly spaced pilot holes in the dadoes in the sides and top, as shown in photo K.

2. Flip the pieces and countersink a 2" screw in each pilot hole from the outside.

Cut the Aprons

The aprons fit between the side pieces at the bottom. Their length is exactly the same as the distance between the inside edges of the stopped dadoes in the top.

1. Rather than using your tape measure, which will require reading and remembering fractions, take a direct measurement using the apron itself. Start by slipping a piece of ¾" scrap into one of the stopped dadoes.

2. Use the scrap as a stop: Place the squared end of a length of ¾" × 2½" apron stock against the stop and let it extend across the top and over the other stopped dado. Mark the location of the inside edge of the other dado on the wood, being sure to put an X on the waste side. Clamp or tape it to a companion piece for the other apron, and cut both pieces to length.

3. Double-check that the length is correct by laying the aprons between the stopped dadoes on the top.

4. Put two pocket holes in each end of both aprons.

Assemble the Case

As always, put the case together with clamps to check both your clamping strategies and to make sure the pieces are correctly made before gluing.

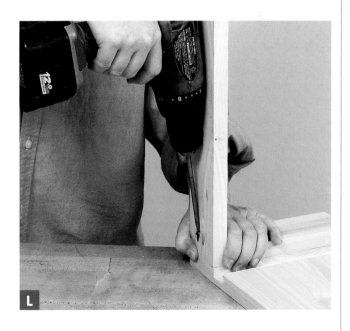

L

M

from the top to the bottom of the side pieces if necessary to hold them in place.

5. Check that the case is square. Use your largest square to make sure the inside angle between the top and the side is 90°. Also make sure the rabbets are flush at the upper back corners. If they're not, make sure the side pieces are all the way into the stopped dadoes.

6. Stand back and take a look at the bookcase from a distance. Look for racking and out of squareness. When everything checks out, remove the clamps and aprons, and sand all the pieces to 220 grit using a random-orbit sander.

Gluing up

1. Apply glue to the ends of the aprons, the end grain of the cleats, the tops of the side pieces, and in the stopped dadoes, as shown in photo N. Replace the clamps.

2. Put the case face down on the bench and measure the diagonals to check squareness before driving the screws.

Dry-fitting

1. Screw the aprons to the front and back of one of the side pieces, as shown in photo L on the previous page. Use your thumb to keep the top edge of the aprons flush with the bottom edge of the dadoes.

2. Lay the assembly on its back edge and fasten the front apron to the other side piece. Use a piece of ¾" scrap in the dado for alignment, as shown in photo M.

3. Gently flip the assembly to rest on the front edges and fasten the back apron in place.

4. Place the top on the bench with the dadoes up, and insert the assembly into the stopped dadoes. Use long panel clamps and clamp

N

3. Hook the tape measure over an upper corner of the case and measure the diagonal to the lower corner, as shown in photo O. Exactly where you hook it isn't as important as being able to hook it in the same relative location when you measure the other diagonal.

4. Measure the other diagonal and compare. If the sides are all the way in the dadoes and the aprons are the correct length, the diagonals will not be very different. If they are, try shoving the bookcase into square and see if that works. If not, loosen the clamps and make larger adjustments.

5. When the diagonals match (or are within about $\frac{1}{16}$"), drive 2" screws into the pilot holes in the top, as shown in photo P.

Install the Shelves

Position the bookcase on the bench with the rabbet side up and the top overhanging the edge so the front is flat on the bench.

Cutting to length

Since the rabbets are the same depth as the dadoes, you can get the length of the shelves by measuring the distance between the rabbets on the back.

1. Push the hook of your tape measure against the rabbet at the upper dado, and open the tape until its case is against the rabbet on the other side, as shown in photo Q. Lock the tape open.

2. Move to the other rabbet and check that the distance is the same. Keep the tape measure locked.

Q

3. Next, transfer the measurement to the workpiece. Hook the tape measure on the squared end of a ¾" × 9¼" shelf piece and put a mark on one edge at the far end of the tape case. Put an X on the waste side.

4. Clamp the two shelf pieces together or use double-sided tape, then crosscut to length.

Checking the fit

The shelves should slide easily into the dadoes. If they don't, deal with them in the same way you dealt with the sides—by chamfering the ends or by planing the faces. Once they're in place, you'll note that even with their fronts resting on the bench, the back edges stick up above the rabbet, making it impossible to fit the back. The shelves must be made narrower by an amount equal to the depth of the rabbet. You could remove this material with a circular saw and guide, but the point of this project is learning to use the router. So, you'll remove the excess using a router and a template-cutting bit.

1. Make a pencil mark at the bottom of the rabbet on both ends of each shelf, as shown in photo R. Remove the shelves from the case.

2. Clamp a shelf to the bench, then clamp a straightedge to the shelves. Align each end of the straightedge with your marks and rout away the excess with the template bit against the straightedge.

Testing your clamping strategy

Gluing the shelves in place requires long panel clamps on each shelf front and back.

1. Keeping the case on the bench with the front side down, insert the bottom shelf and put a clamp across the back. Place the bar right down on the edge, as shown in photo S. Repeat with the other shelf.

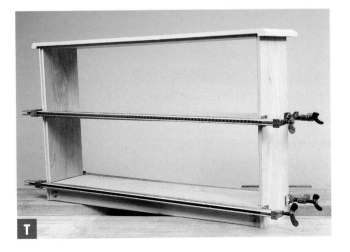

2. Set the case upright, then apply two more clamps across the front, as shown in photo T.

3. Stand back and take a look to see that the sides fit properly. If the sides bow outward at a shelf, it's too long. If they bow inward, the shelf is too short. Adjust as necessary.

4. When the shelves fit, sand both sides smooth using a random-orbit sander and 220-grit disks. Break the edges using sandpaper and a hand-sanding block.

Gluing and screwing the shelves in place

1. Apply glue to the dadoes and to the ends of the shelves and clamp as before.

2. Drive 2" screws through the pilot holes and let the glue dry.

Install the Base Molding

Decorative moldings meet at the corner in non-structural miter joints, as shown in the drawing on p. 226. Measuring and cutting perfect miters around all four sides can be tricky—it requires perfect measuring, perfect cutting, and perfectly adjusted tools. In this project, you'll ace all that by simply applying molding to only three sides. That way, you can concentrate on getting the front corners right.

Before installing any molding, make sure the aprons are flush with the front and back edges of the side pieces. If they stand proud, plane them flush. If they're a little below flush, just leave them alone.

The Elements of a Mitered Corner

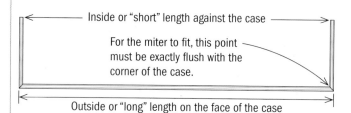

Inside or "short" length against the case

For the miter to fit, this point must be exactly flush with the corner of the case.

Outside or "long" length on the face of the case

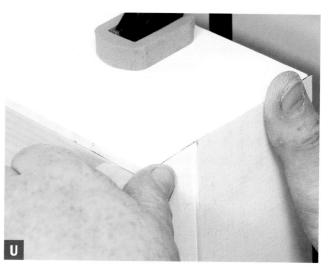

U

Fitting the molding

1. Cut the longest piece first so if you make a mistake, you can use it to make shorter pieces.

2. Cut the miter in one end. Set your miter box or chopsaw to cut a 45° angle and clamp the molding in place with the decorative edge up and the longer end on the face of the molding.

3. Clamp the molding to the front apron at each end. Keep it flush to the bottom of the apron—it looks fine if the shelf is a little higher than the molding. Make sure the inside of the miter (on the back side of the molding) is right at the corner of the case.

4. Draw a line on the back side of the molding along the outside edge of the case. Remove the molding and square the line around both edges.

5. Once more, set up the molding in the miter saw to cut the other miter on the line, with the long side of the molding facing outward, away from the fence.

6. Clamp the molding in place again and check the fit. Leave the clamps in place.

7. Rough-cut a piece of molding about 12" long and miter one end with the long side on the face.

V

8. Hold this piece of molding against the front molding piece that's clamped in place, pushing firmly to get a tight fit at the miters, as shown in photo U. Clamp it in place.

9. Using your square, mark on the molding where to cut it flush with the back edge of the case, as shown in photo V. Set the saw to 90° and make the cut.

10. Repeat with the other corner.

Gluing and nailing the molding

Traditionally, furniture makers fasten molding and trim with tiny brads, although very small finishing nails will also work (see p. 133 for more information). The small heads of brads don't make much more than a dimple in the wood, even when they're countersunk with a nail set. For painted work such as this, you can also use small-headed trim screws and fill the holes with putty.

1. Using 1" brads and a Warrington hammer, start the brad with the straight end of the hammer so you don't mash your fingers, as shown in photo W.

2. When the brad is set in the wood and no longer needs to be held, switch to the face of the hammer. Stop driving just before the head is flush so you don't dent the surface of the bookcase.

W

X

3. Set the brad head below the surface with a nail set by putting the tip of the nail set into the dimple on the brad and driving it slightly below the surface, as shown in photo X. Drive a brad every 6" or so.

Leveling the bottom

When the molding is fastened, flip the case upside down (protect the top from scratches by using a piece of clean plywood, cardboard, or carpet) and plane the bottoms of the side pieces, cleats, and moldings flat so the case will sit level, as shown in photo Y.

Install the Back

If your case is square, cutting the back to fit is a simple job. If something went wrong and the case isn't square, you'll have to cut the back to match.

Cutting the back to fit

1. Once more, set your tape measure for the inside distance between the rabbets at the top of the case. Transfer the measurement to the beadboard by hooking the end of the tape over one edge (perpendicular to the beads) and mark at the back side of the tape case. Repeat for the bottom of the case, resetting the length of the tape if required.

2. Using a circular saw and a 50"-long guide (see p. 195 for more information), cut the piece to width with the guide on your marks.

3. Measure the distance from the rabbet on the top of the case to the bottom edge of the apron and transfer this measurement to the edge of the beadboard, going parallel to the beads. Repeat on the other side of the case and transfer this measurement to the other edge of the beadboard. Cut with your circular saw and guide.

4. Slip the back into place to check the fit. If the fit is very tight, you may have to flex the back a little to get it in. If the back is still too big, use your block plane to shave down one edge. If the back is a little small and some gaps show on the sides, don't worry because you'll have a chance to fill them before painting.

5. Once the back fits, use a straightedge to draw light lines on the back to show you where the shelves lie underneath the beadboard. That way you'll know where to put the nails.

Fastening the back

To fasten the back to the narrow rabbet, use 1" brads or finishing nails rather than screws. Screws are relatively large and would make a mess if you ran through the edge of a shelf. The hole made by a misplaced brad is a much smaller problem, easily filled with a little putty and sanded smooth.

> **WORK SMART**
>
> Check that the front of your shirt and your cuffs are dust free. It's not much good vacuuming everything else if your every movement sends out a cloud of sanding dust.

Z

Finishing

About 80% of getting a great paint job is preparation. When you start with a flat, smooth, clean surface, it's hard to get bad results. But prep work requires a lot of patience. You just have to be disciplined and settle down to doing your best on each rather dull step, even though what you really want is to get the paint on so you can see your finished project.

Filling and sanding

1. Fill all the countersinks, any small gaps in the dadoes that show in the front of the bookcase, knots, dings, and other imperfections with a readily sandable filler (for more information, see p. 144). Apply it with a putty knife or a plastic spreader. Push down hard to get the filler to go into the hole, then scrape off the excess, as shown in photo Z.

2. For any gaps in the back or in the mitered corners of the molding, use some soft, paintable caulk that comes in a tube. It's much easier than standard fillers to mold it into a nice

WORK SMART

Follow the paint maker's directions for solvents, prep, and cleanup. Modern paint chemistry is complex; you can't always mix and match brands or types. Sometimes a paint's chemistry is very specific, and you'll get the best results by using compatible products.

radius with your finger. That way you'll get a smooth surface that won't require sanding.

3. Once the filler is dry, hand-sand using a block and 150-grit paper. When you start, the area around the hole is smeared with a thin layer of filler. Sand the surface until the smear is gone from around the filled hole and the edges of the fill appear crisp, as shown in photo AA on p. 230.

4. As you're sanding, look for holes you missed or areas that need a little more filler. Apply another round of filler, and sand it again.

AA

5. Switch to 220-grit paper and sand a slightly wider area around the filler.

6. Since the rest of the piece was sanded before assembly, simply check it over for smoothness and break all the edges.

Cleaning

Cleaning is the most overlooked step in getting a fine finish. If you neglect it, the painted surface will end up as rough as sandpaper, no matter how much time you spent sanding.

1. Vacuum or brush the bookshelf inside and out. Clean up the surrounding area as well.

2. Wipe the surface with a paper towel wet with denatured alcohol to remove any hand oils or chemical contaminants. This step also picks up any remaining dust.

3. Wipe down the surface with a tack cloth to pick up any remaining dust. If you're using a water-based finish, make sure you have a compatible tack cloth.

Painting

You can use any type of paint you like for this project. Everything looks good on it, from faux-distressed milk paint to super-glossy enamel.

1. For the best finish, start with a primer coat. A primer paint is full bodied and designed to fill the grain, smooth the surface, and sand easily. Check with your paint dealer to find a product compatible with the paint you intend to use. Apply a thin coat or coats as specified by the manufacturer, as shown in photo BB.

2. Sand lightly with 220 grit using a random-orbit sander until smooth to the touch. Don't sand away the paint—remove as little as necessary to get a smooth surface. You won't be able to get right into the corners with the machine, so do that part by hand with the sandpaper on a hand-sanding block. Vacuum and wipe with a tack cloth.

3. Apply a thin coat of top-coat paint. When it's dry, sand lightly with 220-grit paper. You don't want to sand away all the paint, just smooth off the nibs. Vacuum up the dust.

4. To get the surface looking really good, apply at least two coats of paint. If you sand and clean carefully between each coat, the surface will be smooth and even.

> **WORK SMART**
>
> **M**ost paint problems result from applying too much paint, often without allowing enough drying time between coats. Thin coats of paint self-level better, and they dry more quickly and completely.

BB

Storage Bench

The Arts and Crafts movement is often noted for a few simple tenets—honesty of materials, solidity of construction, utility, adaptability to place, and aesthetic effect—all of which can be seen in this storage bench. The bench's straight lines and solid joinery make a strong statement, but adding a couple of seat cushions (available at any home store) transforms it into a soft, comfortable seat. In addition, the top is hinged so you can use the bench to hide any items that pile up next to the back door or at the foot of the bed.

Building projects like this is very much in keeping with the spirit of the piece. The father of the American Arts and Crafts movement, Gustav Stickley, is best known as a furniture maker, but he also published a popular magazine called *The Craftsman*, which aired the philosophy of the movement and was one of the first do-it-yourself magazines. Working in basement and garage workshops, passionate amateurs used plans published in the magazine just as you are using this book today. (By the way, *Mission* is used interchangeably with *Craftsman*, but the term infuriated Stickley, as it was used by one of his competitors.)

If you visit an antiques store, you may observe collectors crawling under tables and chairs as they search for a sticker bearing the Stickley motto: *Als ik kan* (As best I can). But you don't need to buy a $500,000 sideboard to appreciate the value of those three words. The pleasure you feel from working in your shop and the skills you learn when making something yourself is really what it's all about.

What You'll Learn

- Making four-sided quartersawn legs
- Using a biscuit joiner
- Using a slot-cutting router bit
- Cutting tenons with a dado cutter
- Keeping panels square during assembly

Arts and Crafts furniture is distinguished by strong (almost exaggerated) joinery, such as large tenons sticking through even thicker legs. As you learn additional woodworking skills, you may elect to follow Stickley's instructions to the letter, but there are easier ways to get the same effect. Building this storage bench, you'll learn to use a biscuit joiner to quickly connect the rails to the legs. This tool cuts a small slot in both mating pieces, then employs a wooden disk (the biscuit), which works like a small tenon to connect the two pieces. You'll still get a chance to cut a few tenons as you make the stiles, the

Cutting slots with a router. **Your router is good for more than shaping edges. You'll learn to use a slot cutter to make frame-and-panel assemblies, which can be used for making doors and cabinets of any size.**

Keep assemblies square. **This simple assembly table ensures perfect glue-ups.**

Nibbling tenons. **Many woodworkers cut tenons with a table-saw and dado cutter. Here, you'll get to try that technique on for size.**

A Step Beyond Your Basic Bench

Frame-and-panel construction enables you to build big pieces in a small space.
Plywood panels save time and expense without compromising good looks.

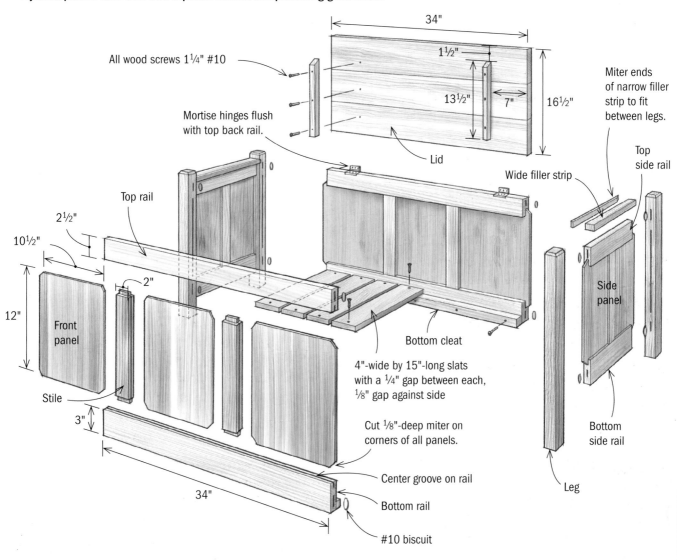

All wood screws 1¼" #10

Mortise hinges flush with top back rail.

34"

1½"

13½"   7"   16½"

Lid

Miter ends of narrow filler strip to fit between legs.

Top side rail

Wide filler strip

Top rail

2½"

10½"

12"

3"

Front panel

Stile

2"

34"

Side panel

Bottom side rail

Leg

Bottom cleat

4"-wide by 15"-long slats with a ¼" gap between each, ⅛" gap against side

Cut ⅛"-deep miter on corners of all panels.

Center groove on rail

Bottom rail

#10 biscuit

vertical pieces of wood that fit between the horizontal rails.

Since the bench is composed of many identical pieces and symmetrical assemblies, this project enables you to think about how projects "flow" through your shop. To minimize measurement-related errors, prepare your stock so that you set your miter saw or table-saw once, then cut all the pieces that are the same size at one time. To ensure that same-size panels turn out that way and a groove cut along one board aligns with its mate, you'll learn how to use spacer boards and jigs that eliminate the need for a tape measure. You'll also learn how to make and use an assembly table that not only frees up space on your bench but also squares up assemblies so that you can concentrate on clamping before the glue sets.

MATERIALS

Quantity	Part	Actual Size	Notes
8	Leg cores	¾" × 1" × 22½"	Quartersawn white oak is preferred throughout this list, but any hardwood will work. Cut the legs long and trim to length after attaching the veneer.
8	Leg veneers	¼" × 1½" × 22½"	Start with 2"-wide strips. Trim to final width after gluing to sides of leg core.
2	Top side rails	¾" × 2½" × 14"	
2	Bottom side rails	¾" × 3" × 14"	
2	Top front and back rails	¾" × 2½" × 34"	
2	Bottom front and back rails	¾" × 3" × 34"	
6	Stiles	¾" × 2" × 12"	
4	Side panels	¼" × 6½" × 12"	You can cut 10 panels (plus a few extra) from a half sheet of ¼" oak plywood.
6	Front and back panels	¼" × 10½" × 12"	
1	Lid	¾" × 16½" × 34"	
2	Lid cleats	¾" × 1" × 13½"	
2	Wide filler strips	¾" × 1" × 14"	
2	Narrow filler strips	¾" × ¼" × 14½"	Cut 1" longer than needed. Miter the ends to fit between the legs after assembling the bench.
2	Bottom cleats	¾" × 1" × 34"	
8	Bottom slats	¾" × 4" × 15"	Pine or poplar is fine, your boots won't know the difference.
12	#10 biscuits		Buy a container.
2	No-mortise hinges	3" × ¾"	See Resources on p. 296.
1 box	1¼" #10 wood screws		
	Polyurethane glue		For laminating legs
	Yellow glue		For gluing biscuits and panels
	150-grit and 220-grit sandpaper		You'll need packs of both sheets and disks.
1 pint	Watco® oil		Dark walnut
2 cans	Spray polyurethane		Satin finish
	Respirator		For use when spraying

Buying Materials

Although many different types of wood were used during the Arts and Crafts period, quartersawn white oak was the wood of choice. Quartersawing produces lumber that is less prone to warping, which is in keeping with two of the movement's guiding principles: strength and permanence. An important side benefit is that quartersawing white oak also reveals a handsome ray fleck figure. You can find quarter-sawn white oak at most local mills, and for a small extra fee they'll mill it to the ¾" thickness you'll need for this project. If white oak is unavailable or prohibitively expensive, consider substituting red oak.

TOOLS

- Tape measure
- Tablesaw with a good rip blade
- Miter saw
- Biscuit joiner
- 12" combination square
- Block and #4 or #5 handplane
- Card scraper
- Bearing-guided chamfer bit
- Router
- ½"-dia. by 1"-high flush-trimming router bit
- ¼"-wide by ¼"-deep slot-cutting router bit
- Clamps
- ⅛"-dia. drill bit
- ⅜" countersink bit
- Drill
- Screwdriver

Building the Bench

This is a relatively large project, but it's not difficult to build—even if your shop is a cramped corner of your garage or basement. The bench breaks down into three bite-size subassemblies: the legs, the panels, and the top. Plan on spending a day to cut and assemble each section and a few hours the following weekend to pull the bench together and apply a finish.

As you build each section, pay close attention to the grain. The fleck pattern on the legs is sure to run in a different direction from the rails, but try to arrange the wood so that the fronts of both front legs match and the top rails complement the bottoms. You'll appreciate the difference this detail makes as soon as you wipe on the first coat of stain.

Making the Legs

The ray-flecked grain of quartersawn wood is a signature of Arts and Crafts furniture. The problem with using a solid leg is that the flecks will appear only on two opposite sides (and one of those sides will be hidden by joinery). There are several solutions, such as mitering the edges of four boards or using a fancy lock-miter bit on your router table. But the simplest fix, described here, is to make the leg slightly undersize in thickness and then glue quartersawn strips to the two straight-grained sides. The glueline between the thin veneer and the leg blank will disappear after you chamfer the edges.

1. Glue up the core portion of the legs first. Start by ripping two 4'-long boards to 2½" wide. Arrange the boards so that the best fleck figure faces out. Wipe a thin coat of polyurethane glue onto the inside face of one board and clamp the other on top. Once the glue has cured, knock off the dried foam with a scraper. Joint one edge (a #4 or #5 handplane works fine), then rip two 1"-wide lengths. Using your miter saw, cut the lengths in half to make four 24"-long core pieces.

2. To make the ¼" veneers that hide the core's glueline, you'll need to resaw a ¾" by 4" by 48" board. (For more on this process, see "Skill Builder: Resawing Wood on a Tablesaw" on pp. 238–239.) Set your rip fence just a hair thicker than ¼" so you'll be able to remove saw marks or burns with a handplane. After resawing, crosscut the veneers to 24" and attach them to the core pieces with polyurethane glue. As you tighten the clamps, make sure the veneers slightly protrude over both edges.

Leg Detail

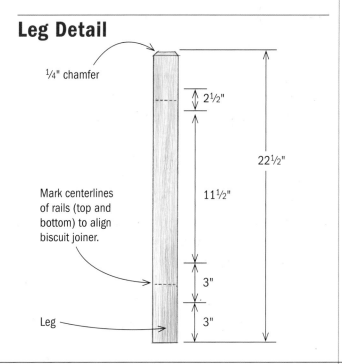

¼" chamfer

2½"

22½"

11½"

Mark centerlines of rails (top and bottom) to align biscuit joiner.

3"

3"

Leg

What You'll Need

- Tablesaw with rip blade
- Auxiliary fence
- 1/8"-thick spacer strips
- 17" to 26" handsaw
- Planer or #5 handplane

A bandsaw is the preferred machine for resawing because it can quickly slice through thick boards without losing a lot of wood to the blade. However, smaller models may not have the horsepower or capacity to handle boards wider than 6". Comparatively, a tablesaw is slower and chews up more wood, but even a small benchtop saw can successfully slice wide boards into thinner stock.

Finish by hand. A handsaw is a safer and more controllable way to finish narrower boards. Insert a wedge in the kerf so the board doesn't pinch the blade.

Slicing two boards from one. Resawing on a tablesaw is slower than with a bandsaw, but with a thin-kerf rip blade, you can slice wide boards easily, even with a smaller saw.

Before resawing, check your stock. A severely bowed or cupped board may be too warped to maneuver safely past the blade. Even if you can manage to resaw warped stock, the final product will require a lot of additional planing to achieve an even thickness.

1. Outfit your tablesaw with an auxiliary fence and a zero-clearance insert with a splitter. (Most saws offer aftermarket inserts that make your tablesaw much safer and are well worth their small price.) Position the fence so the board is centered on the blade. You may also want to use a featherboard to press the board against the fence as you make the cut.

2. Working alternately from both edges, make a series of shallow cuts, about 1" deep, as shown in photo A. Progressively raise the blade until you reach the maximum cutting depth of the saw. By making the cut in steps, you won't stall your saw and you'll find that the cut is easier to control.

Flush-cut edges. Position the bit's bearing against the core.

Plane away the evidence. Use a bench plane to remove burn or blade marks and get the surface as smooth as possible. If you plane off too much, simply adjust the width of the side dado.

Note: Don't use your tablesaw to slice all the way through the board. Cutting through the center can cause the featherboard to press the outside slice against the blade, which will cause burn marks and lead to kickback. Use a handsaw to finish the cut.

3. Clamp the board in your bench vise and insert a few spacers in the sawkerf. Using your handsaw, cut through the web of wood left between the kerfs, as shown in photo B.

4. You can expect some blade marks. When power planing, machine all the boards at the same time to ensure that they are all planed to the same thickness. You can also remove the saw marks with a handplane, a scraper, or 100-grit sandpaper, as shown in photo C.

Chamfering the edges. Set the chamfer bit to the height of the veneer. Chamfering hides the glue joint and softens sharp oak corners.

3. The veneers were cut wider than the core to allow for slippage during glue-up. Once the glue has dried, trim them flush with a router and flush-trimming bit (a bearing-driven straight bit). Knock excess glue off the leg, if necessary, then set the bit so that the bearing rides against the center of the leg (see photo A). Remember to feed the stock from right to left, against the rotation of the bit.

4. Using the router table and a 45° bearing-driven chamfer bit, raise the bit so that the top edge of the chamfer just touches the glueline, then rout all four edges of each leg, as shown in photo B on p. 239. When finished, the glueline will be unseen because it's aligned with the edge of the chamfer.

5. Trim the four legs to 22½" long at the miter saw. Next, set the saw to cut a 45° bevel and cut a ¼"-tall chamfer on the top ends of all four legs. To make an even chamfer on all four sides, clamp a stop block to your saw's fence so all you need to do is rotate the leg and make the cut.

6. The best way to avoid measurement-related layout mistakes is to mark out the locations of the top and bottom rails on all four legs at once. Position the bottoms of the legs against a straightedge, such as your framing square, and use your combination square, as shown in photo C. Mark the locations of the top and bottom edges of the rails, then mark the centerlines for both rails. You'll use these reference lines to cut the biscuit slots.

Wraparound Quartersawn Legs

To make square legs—even if your wood isn't exactly as thick as the dimensions in the drawing—adjust the width of the core so that it's equal to the leg thickness minus twice the thickness of the veneer.

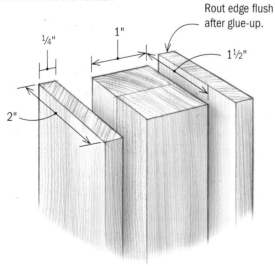

Measure once, then mark all four at once. When marked off in one swipe, the legs won't be off even the width of a pencil line.

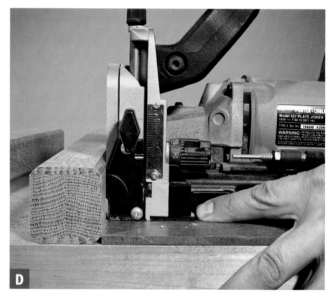

Biscuit the legs. To create a ¼" offset, position the leg so the outside face is against the benchtop and insert a hardboard spacer beneath your biscuit jointer as shown.

7. The rails are offset from the inside edges of the legs by ¼". The simplest way to cut matching biscuit slots is to use a ¼"-thick spacer instead of the biscuit joiner's fence to slot the legs. Using your benchtop as the reference surface, position the leg so that the inside face is against your workbench. Place the biscuit joiner on top of the spacer, then align the centering mark on the biscuit joiner with the centerline

How a board is cut from a log can make a big difference in the appearance of the wood. Most boards are flatsawn—the log is sliced tangentially to the annual growth rings—a method that produces an arched or "cathedral" grain pattern on the faces. Flatsawing is fast and produces the most usable wood from a given log, but flatsawn boards have a tendency to cup. In some projects, like this one, the arched grain can be distracting unless carefully arranged.

Quartersawn boards are cut so that the annual rings are perpendicular to the face of the board. Quartersawing involves more work (the log is first cut into quarters and reoriented before every cut so that the blade is perpendicular to the annual rings) and produces narrower boards than flatsawing, but the straight-grained stock is stronger and less likely to warp. In some species (oak especially), it also produces large cross-grain

flecks, which are a favorite with builders of Arts and Crafts furniture.

The quartersawing process doesn't always produce boards with rings perpendicular to the face. When the annual rings run from 30° to 60° to the face of the wood, it's considered riftsawn. Riftsawn boards are straight grained and are more dimensionally stable than flatsawn, but most of them lack the bold fleck figure found on true quartersawn stock.

on the leg, and cut the slot for the top rail, as shown in photo D. After cutting the slot for the bottom rail, flip the leg end over end so that the inside face of the leg is facedown on the bench before cutting the slots for the adjacent corner.

Building the Sides

The next steps cover the assembly process for the two side panels. But at the same time, you will also cut many of the pieces you'll need to build the front and back panels. Making all the same-size cuts at once saves time and reduces errors.

1. Set your tablesaw fence to 3" and rip enough wood to make two 14"-long bottom side rails and two 34"-long bottom front and bottom rear rails. Reset the fence to 2½" and cut the four matching top rails. Next, set the fence to 2" and

rip enough wood for the six 12"-long stiles that will be used for the side, front, and back assemblies. Using your miter saw or crosscut sled, cut the rails and stiles to final length.

2. Cut the plywood panels next. Set your rip fence to 12". Orient the half sheet of ¼"-thick plywood so that the grain on the good (top) face is running perpendicular to the blade and cut three strips.

3. Use the crosscut sled to cut four 6½"-wide panels for the sides and six 10½"-wide panels for the front and back. You may want to cut a few extra panels with the leftover wood just in case you notice an ugly glueline in the veneer or mismatched grain later on. To allow for the radius left in the inside corners by the slot cutter router bit, set your miter saw to 45° and nip ⅛" off the corners of the panels.

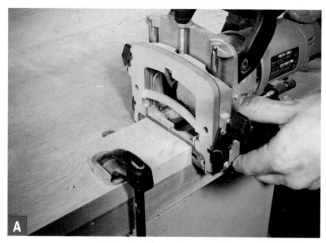

A

Biscuit the rails. Lay the "good" face of the rail against your bench and cut a slot without a spacer. By doing this, the biscuit connecting it to the leg will automatically establish the ¼" offset.

B

Measurements not required. Cutting two spacers from scrap to establish the height of the legs and panel height is faster and just as accurate as using a tape measure and triangle.

4. Take a close look at your rail pieces and mark the face with the least attractive grain as the back. Next, lay the back face of the rail against your bench, mark the centerline, and use your biscuit joiner to cut slots in both ends, as shown in photo A. (The slot won't be centered exactly on the thickness of your stock. Using a spacer to raise the slot on the leg and no spacer when slotting the rails automatically establishes a stepped or offset joint.)

5. With the back face of the side assembly facing up, dry-assemble the side rails to the legs. (Because biscuits are designed to allow some "wiggle room," it's easy to accidentally assemble a panel out of square. To prevent this, use the self-squaring assembly table shown in photo B.)

Cut grooves to accept the panels

Because this may be your first frame-and-panel case piece, the joinery process is designed to enable you to see how everything goes together even before you finish cutting all the parts. You'll dry-assemble the (panel-less) sides of the case, then use your router and slot-cutting bit to groove the panels and stiles. Cutting the slot this way prevents the chance of accidentally grooving the wrong edge, stops the slot cutter from cutting all the way down the leg, and ensures that the grooves line up on all four sides. Just as you did when cutting the biscuit slots, you'll use spacers throughout the process to ensure accurate alignment—without having to adjust the depth of your router or even use a ruler.

1. Dry-assemble one side of the bench so that the outside (good) face is against your benchtop. You'll notice that the inside faces of the rails are offset from the legs by ¼". To provide a level surface for your router, temporarily attach two ¼"-thick hardboard spacers to the rails with carpet tape.

2. A ¼"-wide by ¼"-deep slot-cutting bit is basically a miniature sawblade. Grooving with a router, instead of your tablesaw, enables you to cut the panel grooves in the legs and rails in one step. Adjust the height of the bit so the slot is roughly centered on the edge of the rail. Rest the base of the router on the side assembly and

What You'll Need

- ¾" by 4' by 4' plywood
- ¾" by 3" by 2' wood or wood strips
- Two sawhorses
- Framing square
- Drill with bits and driver
- 1¼" deck screws
- Foam brush
- Polyurethane
- Wax

This fixture is about as simple as it gets—all it takes is a half sheet of plywood or particleboard, a few scraps of wood, and a pair of sawhorses—but once built, it's likely to become a permanent fixture in your shop. The two built-in squaring strips serve as a reliable reference point when assembling doors or panels. Simply position one side of your project against

one strip, then clamp the other against the second strip to establish a square corner. The squaring guide is useful when assembling doors, the side panels of this bench, or any square assembly, but for those times when you need a smooth, flat surface, just flip the top.

1. Start with a half sheet of ¾"-thick plywood or particleboard. Apply two to three coats of polyurethane, making sure to thoroughly seal the ends.

2. Using a framing square as a guide, attach two strips of wood in an L-form with 1¼" deck screws. If you plan to build doors, leave a gap along the inside corner so that the rails can stick through, as shown. (Consider these guides sacrificial; you trim them as needed to fit a clamp or work around a protruding tenon.)

3. Apply a heavy coat of paste wax to the top and the guides. Apply another coat of wax whenever it takes more than a light pass with a scraper to pop off dried paint or glue.

A Self-Guided Assembly Table

Screwing a pair of guide strips to your assembly table ensures that at least one corner of your glue-up is square. It's especially useful when building frame-and-panel assemblies.

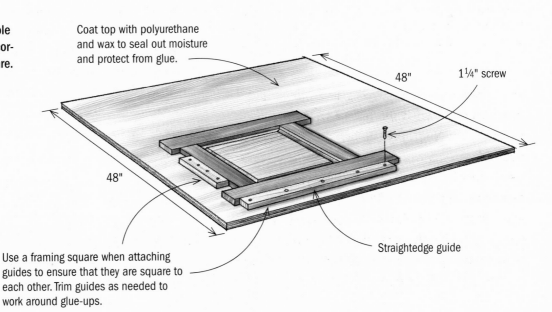

Coat top with polyurethane and wax to seal out moisture and protect from glue.

48"

1¼" screw

48"

Straightedge guide

Use a framing square when attaching guides to ensure that they are square to each other. Trim guides as needed to work around glue-ups.

Simply slotting. **Center the slot cutter on the middle of the rail, then slot the side and leg to accept the plywood panel. Attach hardboard strips to the leg to compensate for the offset.**

Routing long rails. **It's easier to rout the long rails on your bench. Position the piece so the good side faces up, clamp a ¼"-thick hardboard spacer, then rout the slot from left to right.**

feed it around the inside of the frame in a clockwise direction, as shown in photo C. Repeat the same process on the other side panel.

3. After routing the grooves for the two side panels, the next step is to groove the legs for the front and back panels. To do this, use your side rails as shorter "stand-ins." Disassemble the side panels and rotate the legs so that the short (side) rails are positioned where the longer (front and back) rails will be and clamp them in place. Remember that the back faces of the legs and rails are still facing up. Attach the ¼"-thick hardboard spacers to the rails and rout the legs, just as before.

WORK SMART

If you're planning to stain this piece, sand and stain the panels before assembly. Prestaining will also prevent any light-colored splotches that happen when glue squeezes out of the joints and seals the surface of the wood.

Stile-slotting jig. **This simple setup keeps clamps out of your way so that you can slot the stiles in one pass.**

Dado-cutting short tenons. **Nibble out a ¼"-long stub tenon on both ends of the stiles. Use the groove you routed in the edges as a guide.**

4. To rout panel grooves in the front and back rails, position the boards on your bench with the good faces down. Again, use a ¼"-thick hardboard spacer between the router and rail as you rout the groove, as shown in photo D. You will need to stop the router and reposition your clamps in order to finish cutting the groove.

5. The edges of the stiles must also be grooved to accept the panels. Short pieces can be tough to clamp, so you may want to build a grooving jig, like the one shown in the drawing below. Position the stile in the jig so that the back face is oriented up. As you rout the groove, as shown in photo E, remember to run the router from left to right, against the rotation of the bit.

6. The end of each stile has a small tenon that fits into the grooved rails just like the plywood panels. To cut this tenon, replace your table-saw's blade with a ½"-wide dado cutter. Adjust the cutter's height so it comes up just to the height of the groove. Next, slide the rip fence so that it's ½" away from the inside edge of the cutter. Use your saw's miter gauge to guide the stile as you make the cut, as shown in photo F. In the event that your slot isn't perfectly centered on the stile, cut the tenon on the front or back face of all the stiles, test the fit, then double-check the height of the cutter before cutting the opposite faces.

Stile-Grooving Jig

Made from scraps of plywood, this simple jig will hold the stiles in place as you rout the grooves.

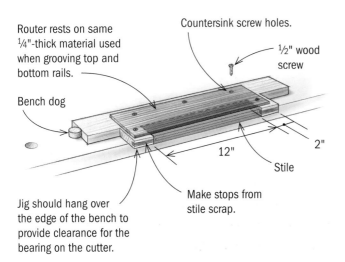

Router rests on same ¼"-thick material used when grooving top and bottom rails.

Bench dog

Jig should hang over the edge of the bench to provide clearance for the bearing on the cutter.

Countersink screw holes.

½" wood screw

12"

2"

Stile

Make stops from stile scrap.

Assembling the Box

To reduce the number of clamps needed, and to eliminate the insanity of a complex glue-up, this bench is glued together in steps: The sides are assembled first, then the front and back, and then the box is joined together. With time on your side, check each subassembly carefully to make sure that everything stays square as it's clamped together. During this process, you'll learn techniques to keep the pieces perpendicular so you can focus on juggling clamps and wiping away glue drips with a damp rag before they dry into hard, unstainable lumps.

1. Start with the side panels. Position the legs, panels, and short rails on your assembly table so the good sides face up. Install biscuits and glue, use the 3"-wide spacer to set the height of the bottom rail, then insert the panel and stile. Once the pieces are assembled and clamped together, use a drafting triangle to double-check that everything is square, as shown in photo A on p. 246. If not, reposition the clamps to make adjustments.

2. Assemble the front and back panels next. To align the ends of the top and bottom rails, you'll need to make a pair of grooved clamp boards. The clamp boards will take the place of the legs during this step of glue-up. Starting with two pieces of ¾"-thick by 16"-long scrap, rout a ¼"-wide by ¼"-deep groove along one edge, as shown in photo B on p. 246. The plywood should fit into the groove, enabling you to butt the ends of both rails against the edge of the caul wood (see photo B inset).

3. Once the glue has cured, carefully inspect the front and side panels for glue stains, black stains (left by pipe clamps), or uneven spots where the rails meet the stiles. Sand, scrape, or plane away these imperfections, as shown in photo C on p. 246, then finish-sand the legs and rails up through 220 grit.

Pulling it all together. Pay attention to the distance between the outside edges of the rails and the ends of the legs. The biscuits may slide a little as you tighten the clamps.

Grooved clamping caul. Cutting a ½"-deep groove on the cauls allows you to align the ends of the rails and establishes the length of the plywood tongue.

WORK SMART

Remember to use yellow or hide glue when using biscuits. Biscuits are made from compressed wood. They need the moisture found in those glues in order to swell and tighten up the joint.

Scrape the stiles. A scraper is useful for removing glue drips and flattening any ledge that might exist between the stiles and rails.

4. It's finally time to assemble your bench. Position a side assembly so that the inside face is up. Insert two scrap 2×4s under the side to provide clearance for the clamps. As you glue and clamp the front panel in place, use a 3"-wide (leg) spacer to make sure that the lower rails of all four sides are the same distance from the bottom of the leg. As you remove the clamps to attach the back panel, keep a plywood triangle in place for support, as shown in photo D on the facing page.

Simple braces help keep the overall assembly square as you pull everything together at glue-up. These squaring jigs can be made simply by mitering the corners off any scrap of ¾"-thick plywood or MDF. (Factory-cut edges are about as square as they come, but if you're not sure if you or the factory made the cut, check the outside corner with a drafting triangle or framing square to be sure.) Drill holes through the outside corners so that you can use them to hold your clamps. You may also want to nip the inside corner so that it doesn't get in the way of (or accidentally become glued to) the pieces you glue together.

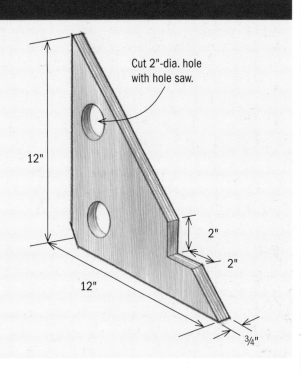

Cut 2"-dia. hole with hole saw.

12"

12"

2"

2"

¾"

D Triangles serve as an extra pair of hands. **The squaring jigs prevent the back from tipping as you tighten the panel clamps. They also ensure that corners stay square as the glue dries.**

E Clamp up the case. **Insert wood spacers under the bench to allow clearance for the clamps. Positioning triangles in opposite corners helps keep things square.**

5. Now that both sides are glued into one side panel, you'll need to glue the front and back to the remaining side in one step. Use the squaring braces to ensure that the bench stays square as you tighten the clamps, as shown in photo E above. If opposing corners line up with the plywood triangles, you can be certain the case is square.

Finish out the inside

1. To secure the bottom of the bench, you'll use two cleats attached to the insides of the front and back assemblies. Cut the two bottom cleats to size, then drill and countersink the holes needed to screw them to the front and back bottom rails, as shown in photo F.

2. Double-check the inside dimension of the box before cutting the bottom slats to size. Before installing the slats, chamfer the top edges with either a router or a handplane. Next, position the slats on the bottom cleats inside the bench. There should be a ¼" gap between the boards. Exact spacing is not critical, but make sure the gap looks even before you screw the boards in place, as shown in photo G.

3. The lid sits between the legs. To conceal the edge of the top rails, I used filler strips cut to fit between the legs. The 1¼"-wide filler strip looks like a single piece of wood, but it's made up of two strips, a wider piece that fits between the flat portion of the legs, and a narrow piece that's mitered to fit between the chamfers cut on the inside corners, as shown in photo H on the facing page. Start by making two 2¾"-wide by 15"-long strips (one for each side of the top).

> ### WORK SMART
>
> **W**hen attaching the slats, start at both ends and work toward the center. If your spacing is a little off, try adjusting the spacing between boards or planing off a small amount from a few of the center boards.

A cleat to hold on to. Screw a pair of cleats to the bottom edge of the lower front and back rails. Use clamps to hold the cleat flush to the bottom edge as you screw it in place.

Plank by plank. As you install the planked bottom, start from the outside ends and work inward.

H

Two strips are easier than one. **Cut the square and miter end from two different strips, then glue them together.**

I

Miter reminder. **Use a thick pencil line as a reminder not to flip the strip the wrong way when you bring it to the saw. Mark the actual cut with a knife or sharp pencil.**

Rip a ¾" strip off both boards. Next cut the two wider boards so that they fit between the front and back legs. Rip the boards so the edges of the wide filler boards touch the inside edges of the chamfers on the legs.

4. Set your miter saw to 45° and miter one end of the thinner filler strip. Placing it against the wider strip you just made, butt the mitered end against one chamfer and mark the opposite end, as shown in photo I. Miter the opposite end, then glue the two pieces together. The filler strip assembly is then simply glued to the top edge of the side rail.

Making the Lid

When building a wide panel, as for a tabletop or lid, spelling out exact dimensions of each board doesn't work well in the real world. You will need to look at your wood as you refer to the materials list. Depending on what's available, you may be able to assemble the top from two boards, or you may need three or more. If you pay attention to the grain and strive for tight-fitting gluelines, either choice can be equally attractive.

1. Edge-join as many boards as necessary to make a panel that is slightly larger than the final 16½" × 34" lid. Trim the lid to the final dimensions after glue-up. (If the lid is too wide for your crosscut sled, you may need to cut it with a circular saw.)

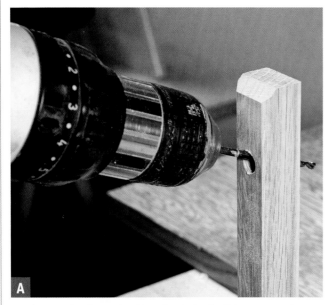

Room to grow. Screwing the cleats to the lid prevents the panel from cupping, but widening the outer cleat holes allows the lid to move with seasonal changes in humidity.

Mortise the upper back rail. Mark their location with a utility knife, then chisel out each mortise by hand.

2. Cut the two lid cleats to size, then drill and countersink the screw holes used to attach the cleats to the lid. The holes on the far ends of each cleat should be wider to allow the screw to pivot as the lid moves with seasonal changes in humidity. To make these elongated holes, flip the board so that the cleat is facing lid edge up, and drill a ⅝"-diameter hole up to the counter-sunk hole you made on the opposite edge. Next, redrill the pilot hole, but tilt the drill to elongate the center of the hole, as shown in photo A.

3. Place the lid facedown on your bench and attach the cleats with six 1¼" screws.

4. The two leaves of a no-mortise hinge are designed to fold into each other, minimizing but not completely eliminating the thickness of the metal hinge. To make the lid sit flush, the hinge should be mortised into the top rail. Position the hinge on the rail and use a marking knife to lay out your mortise lines (see photo B). Next, use a marking gauge, set to the thickness of the hinge, to establish the depth of the mortise. Make a series of shallow cuts with a dovetail saw, then use a chisel to pare out the waste.

Finishing

The best way to get the plywood panels to match the solid-wood rails and stiles is to use a dark stain. A pigmented oil varnish can be used as a finish by itself, but you'll probably want the extra stain and scuff protection offered by a polyurethane varnish. As a side benefit, alkyd-based polyurethanes add a slight amber tint to the wood, which more closely resembles the varnish used on many original Arts and Crafts pieces.

Craftsman makeover. Dark walnut stain approximates Stickley's trademark finish. More important, it helps the red oak plywood blend with the white oak.

Quick spray finish. Perhaps it's not as thick as a brush-on finish, but spray polyurethane offers protection and extra color.

1. Because the legs and panels have already been cleaned up, you shouldn't have much sanding to do. If necessary, use a random-orbit sander to sand the surfaces to 220 grit. (If you're hand-sanding, be careful not to introduce cross-grain scratches where the stiles meet the rails.)

WORK SMART

If the lid of your bench keeps getting bumped against the wall or is dropped in the open position, it can damage the wood, the hinge, or both. To prevent this, consider attaching a small leather strap from one of the side rails to the top of the lid. Make the strap long enough so that the lid can open just past 90°.

2. After sanding, apply a coat of the pigmented oil finish with a rag, as shown in photo A. Depending on the wood, you may need to apply a second coat on some spots to even out the color.

3. Allow two or three days for the oil finish to cure, then apply two coats of polyurethane. If you don't mind the fumes, the fastest way to do this is with a spray can. Brushing would leave a thicker film, but it may also leave drips in the corners of the piece. As you spray, keep the can parallel to the workpiece and a steady 6" to 12" away from the surface (see photo B). Overlap each painted swath about 50% with the next until the entire surface is covered.

4. After allowing a day for the polyurethane to dry, you're ready to reattach the lid to the bench. If you like the sheen and feel of a wax top coat, you can apply a light coat of wax. Immediately after that, you can throw a few cushions on top and have a seat.

Serving Table

This table was designed for kitchens big and small. With a sturdy base and 35¼"-high top—a comfortable height for use as a work surface—it is perfectly suited for chopping, mixing, or other food-preparation chores. It's also attractive enough to use as an extra serving table for drinks or hors d'oeuvres when entertaining a large crowd. At the other extreme, a couple struggling with a cramped kitchen could buy a pair of standard bar stools and use this dining table for everyday meals. With a compact 23¾" by 36" footprint, it would also be a nice addition in a sunroom or other small nook.

If this table looks Shaker-esque, it's because it was inspired by the drawings of several different Shaker tables in John Kassay's *The Book of Shaker Furniture* (University of Massachusetts Press, 1980). The clean lines and functionality of the piece are certainly in keeping with the basic Shaker tradition, but the project's beveled top gives it a slightly modern flair. No doubt, true Shakers would be aghast at the sight of a bottom tray designed as a catchall for kitchen miscellany. So much for their notion of "a place for everything, everything in its place."

The mortise-and-tenon joinery for the base and the dovetailed tray below offer the perfect opportunity to practice your hand-tool skills and enjoy a few quiet evenings in the workshop. Most of the joinery can be accomplished with a few basic machines, but honing your favorite plane and a few chisels will make it easier to sneak up on perfectly fitting joints.

What You'll Learn

- Cutting mortises
- Cutting tenons on the tablesaw
- Routing curves with a template
- Cutting tapers on the tablesaw
- Cutting dovetail joinery
- Rubbing out a finish

The joinery for this table looks more difficult than it really is. For example, the base assembly employs mortise-and-tenon joints to lock everything together. You'll learn how to chop a mortise using a drill press and chisel, but if you don't feel up to that task just yet, you'll also learn how to use your tablesaw to tackle the job.

The bottom tray makes use of the quintessential joint of fine woodworking: the dovetail. Don't let the name scare you. Cutting the single-tail joint boils down to making a few short, straight cuts—most of them on your tablesaw. Of course, you can simply rabbet the ends of the tray sides and lock the corners together with dowels and glue.

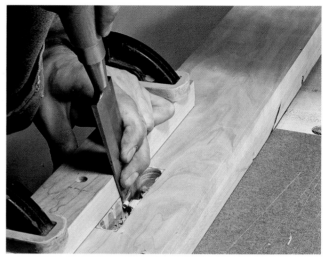

Chop to it. Drilling out most of the mortise and paring up to the line is easy to master, and easier on your chisel than chopping with a mallet.

This is a more involved project that will require a little more shop time, but you'll learn skills and build a few jigs that will make future projects easier and safer. The edge guide will help you accurately trim this tabletop with a circular saw, but it will also allow you to zip through future long cuts. This jig offers a second edge for use with your router and favorite straight bit. The tenoning jig is another useful project. You'll appreciate the way it hops onto your tablesaw's fence and quickly cuts cheeks in just two passes. Last but not least, you'll learn how to make featherboards to help hold boards closer to blades and bits than fingers should ever go.

When it comes time for finishing, you'll learn how to apply varnish and rub out a finish. Unlike wipe-ons, brushed-on varnish creates a thick film finish that provides better protection from water and food stains. Rubbing out (sanding with progressively finer grits) erases the bumps and bubbles that happen every time you pick up a brush, and it leaves a finish that's showroom smooth.

Tablesawn tails. By adjusting the angle of your sawblade and using a tenoning jig, you can use your tablesaw to make the trickiest cuts of the dovetailed tray.

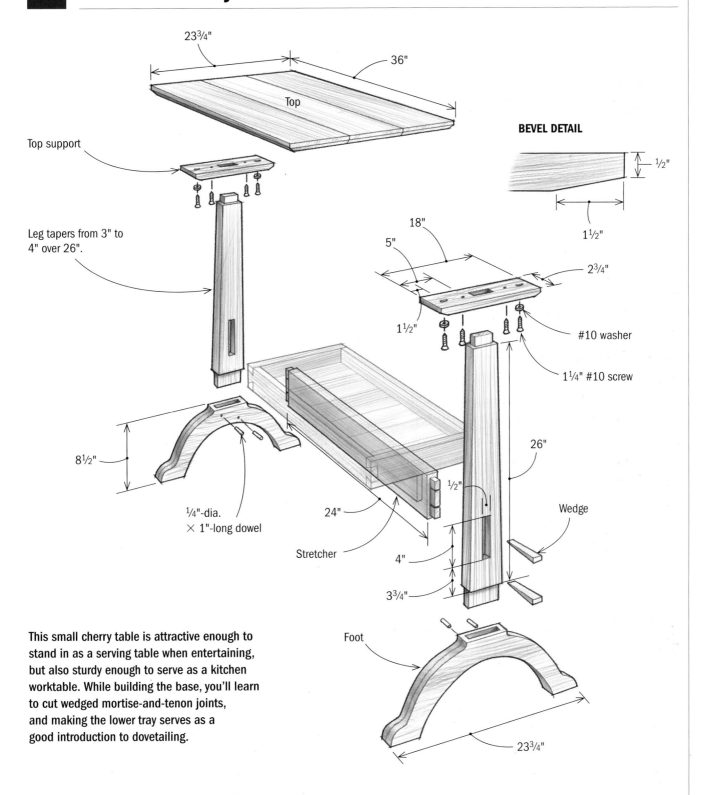

23¾"

36"

Top

BEVEL DETAIL

½"

1½"

Top support

Leg tapers from 3" to 4" over 26".

18"

5"

2¾"

1½"

#10 washer

1¼" #10 screw

26"

8½"

¼"-dia. × 1"-long dowel

24"

½"

Wedge

Stretcher

4"

3¾"

Foot

23¾"

This small cherry table is attractive enough to stand in as a serving table when entertaining, but also sturdy enough to serve as a kitchen worktable. While building the base, you'll learn to cut wedged mortise-and-tenon joints, and making the lower tray serves as a good introduction to dovetailing.

MATERIALS

Quantity	Part	Actual Size	Notes
1	Top	¾" × 23¾" × 36"	All wood is cherry (unless otherwise noted), but a similar hardwood would work as well. You may need three or four boards to make a 23¾"-wide top.
4	Legs	¾" × 4½" × 30"	Face-glue two pieces to make each 1½"-thick leg.
4	Feet	¾" × 8½" × 22"	Laminate two pieces together to make each 1½"-thick foot.
1	Stretcher	¾" × 5" × 27"	
4	Wedges	½" wide × 3" long	Cut wedges from a contrasting wood.
4 pieces	Dowel	¼" dia. × 1"	Can be of the same wood or a contrasting species
2	Top supports	¾" × 2¾" × 18"	
2	Tray sides	¾" × 3¼" × 20"	
2	Tray ends	¾" × 3¼" × 16"	
1	Tray bottom	¼" × 3¼" × 15½"	Consider using a different wood to contrast sides of tray.
2	Tray supports	¾" × 3½" × 20"	
1 bottle	Polyurethane glue		Foam helps hide small gaps, and dried polyurethane glue is easier to knock off than yellow glue.
1	Natural-bristle brush		Buy a high-quality brush and take the time to clean it well.
2 pieces	Felt		To pad inside face of tray supports
1 quart	Varnish		See Resources on p. 296.
1 quart	Varnish thinner		For thinning first coat of varnish and cleaning brush
	Miscellaneous		120-, 180-, 220-, 320-grit sanding disks (2 of each), 1 sheet 320-grit sandpaper, 1 sheet 600-grit wet-or-dry sandpaper, tack cloth, 0000 steel wool, spray adhesive

Buying Materials

You could build this table from the ¾"-thick hardwood from your home center, but this project offers a good opportunity to visit your local hardwood supplier. Realize that wood at a mill isn't usually ripped in standard widths but is sold by the board foot. A board foot is the amount of wood contained in an unfinished board 1" thick, 12" long, and 12" wide. This project requires about 20 bd. ft. of material, but you should plan on buying about 50% more. When you finish building this project, you can use the extra material for smaller projects.

You'll need to glue up a few narrower boards to make the top, so board width isn't a concern, but you'll want to pick up at least one 9"-wide by 8'-long board (that's about 6 bd. ft.) to build the feet.

TOOLS

- Tape measure
- Tablesaw with a good rip blade
- Circular saw with a 40-tooth crosscut blade
- Dovetail saw
- Marking knife
- Marking gauge
- Chisel
- Miter saw
- Jigsaw or coping saw
- Sanding drum
- Mill file
- Drill press (optional)
- 12" combination square
- Block and #4 or #5 handplane
- Shoulder plane (optional)
- Card scraper
- Planer (optional)
- Router and router table
- Chamfer and flush-trimming router bits
- Clamps
- ⅛" and ⅜" brad-point drills
- ½" and ¾" Forstner bits
- Drill
- Screwdriver
- Random-orbit sander
- Hand-sanding block (wood block will do)
- 2' level

Building the Table

Like the earlier projects in this book, you can manage this big project by separating it into parts—the top, legs, feet, stretchers, and tray—and finishing each step before focusing on the next. However, you need to think about the "big picture" first. After milling all stock to ¾", lay your boards across your bench (or between two sawhorses) and decide which boards will be used for each part. Pick out the best boards for the top and the next best for the legs and feet. You'll cut the top supports, stretcher, and tray from what's left over. Mark the wood with chalk or a carpenter's pencil to help you remember your choices.

Gluing Up the Top

Because the top is the focal piece of this project, this is not the place to be frugal when selecting wood. Cutting off or avoiding knots or sapwood will cost you a few extra dollars, but it's better than having to stare at a blemish with each morning's cup of coffee.

The top is 23¾" wide to accommodate the maximum cutting width of most benchtop saws. If your saw has a wider cutting capacity, consider bumping up the width to 24".

1. Working from your best boards, choose enough stock to make the top a little wider than the finished dimension—aim for about 27"—and cut three or four pieces to 38" long. The extra length will allow you to stagger the boards a little when arranging them for glue-up.

2. Joint one edge of each board with a hand-plane or on the tablesaw, then rip and joint the opposite edge. (You still want to make a panel

A

Hammered into alignment. You can expect boards in a glued-up panel to shift somewhat as they're clamped tight. With a hammer and scrap block, a few taps can pull the edges flush (before the glue sets).

B

Scored then sawn. When a cut has to count, score the good face with a knife to prevent the blade from splintering the surface. Clamp both ends of the edge-cutting guide to ensure that neither it nor your work shifts in midcut.

that's 1" or so wider than your finished top.) Dry-clamp the boards before gluing to ensure that there are no gaps along the joints. If everything looks good, you're ready to glue and clamp, as shown in photo A.

3. Once the glue is dry, trim the top to its final dimensions. First, rip one side on your table-saw to get a clean edge. (If you're using a bench-top saw, this should be about 24".) Next, reset the fence to 23¾" and rip the opposite edge to final width.

4. The top is too wide to crosscut on a tablesaw. Instead, use your circular saw and a straight-edge guide. To reduce the risk of splintering along the edge, score the cut line with a utility knife, as shown in photo B on the previous page before cutting. And consider investing in a fresh blade for your circular saw. A 40-tooth thin-kerf crosscut blade is good for most wood-working tasks.

5. Now that the top is trimmed to size, the next step is beveling the bottom edge. If you haven't already done so, you'll need to increase the height of your rip fence, as shown in "Making an Auxiliary Fence" on the facing page. You may also want to make a few featherboards to help press the stock against the fence while it's fed

WORK SMART

If you have a separate, dust-free finishing room, you might want to start finishing the top before completing the base. For extra protection, you'll want the top to have a few more coats of varnish than the base. If you start early, both parts will be ready for a final rubdown at the same time.

D

Bye-bye to burn marks. **Blade marks are unavoidable and easy to deal with. Set your favorite plane to make a light cut and balance the sole on the bevel.**

past the blade. To learn how to make your own featherboards, see "Skill Builder: Making and Using Featherboards" on pp. 260–261.

Tilt your sawblade to 15°, then set your fence so that it's ⅝" away from the bottom edge of the blade (just shy of your finished profile). Use a scrap board to test the cut before cutting the top. Bevel the ends first, then the sides, as shown in photo C.

6. Use a handplane to remove any burns or marks left by the blade, as shown in photo D. Alternately, you can use sandpaper; just be careful not to round over the edge.

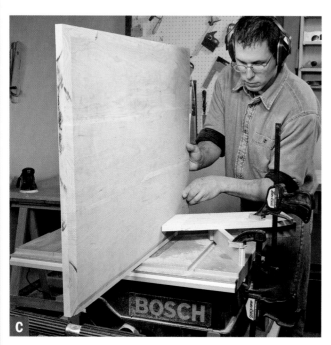

C

Balancing act. **Using a tall auxiliary fence and a featherboard makes this cut safe and easier than it looks. Position a roller stand or sawhorse to support the top at the end of the cut.**

When resawing wide boards or cutting the beveled edge on raised panels at the tablesaw, you'll need to balance boards on their ends and edges. Most fences are too short to safely support the boards during these operations. Attached to your saw's existing fence, an auxiliary fence provides the extra height you'll need to accurately guide the board past the blade.

Not all saw fences are created equal. Some fences have grooves that can accept T-head bolts through the top of the fence. If that's the case with your saw, simply drill a hole in the support board and attach it to the plywood fence. But if your fence has a smooth top, you'll need to rip a spacer strip exactly the same width as the fence so the auxiliary fence straddles your rip fence snugly, as shown below.

T-BOLT AUXILIARY FENCE

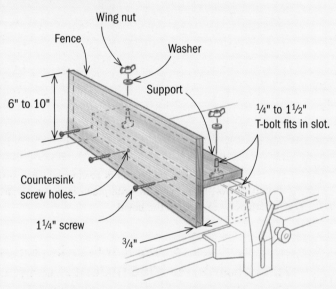

Wing nut
Fence
Washer
Support
6" to 10"
¼" to 1½" T-bolt fits in slot.
Countersink screw holes.
1¼" screw
¾"

Different fences call for different designs. A tall auxiliary fence prevents tall boards from tipping and can protect the metal fence when the dado cutter is set to make close cuts. Unlike some other jigs, you can keep the fence in place even when making regular rip cuts.

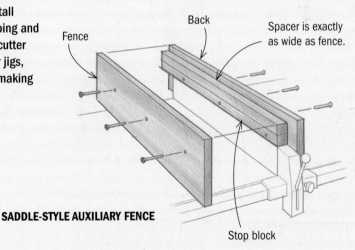

Fence
Back
Spacer is exactly as wide as fence.
Stop block

SADDLE-STYLE AUXILIARY FENCE

SKILL BUILDER: Making and Using Featherboards

What You'll Need

- ■ ¾" by 4" by 16" piece of plywood or MDF
- ■ Small piece of hardwood
- ■ Glue
- ■ Straight-grained boards, at least 2" wide by 6" long (longer is better)

Featherboards, or fingerboards, are simple jigs that excel at holding narrow or thin stock securely against a table or fence without putting your own fingers at risk. They also prevent kickback and help produce smooth, consistent cuts. Almost any setup in which a board is guided past a fence can be both safer and more accurate with a well-placed featherboard.

You can make featherboards in a variety of shapes and sizes.

1. Set your sawblade to 60°. Because the bevel gauges on the front of most saws can be difficult to read or unreliable, consider using a small drafting triangle to set the blade, as shown in photo A.

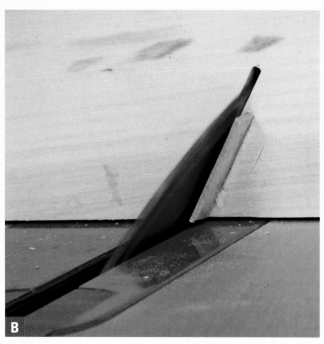

B

Stronger or springier fingers. **The distance from the pin to the cut will determine the spacing between fingers and the amount of give in your featherboard. Start with ⅛" spacing, then adjust to suit your preference.**

A

Setting the blade. **Set the triangle between the carbide teeth and adjust the bevel angle until the blade rests against the plastic triangle.**

Featherboard-Making Jig

Clamp this two-board jig to your miter gauge and make your shop a little safer. To adjust the resistance, change the height or spacing of the "feathers."

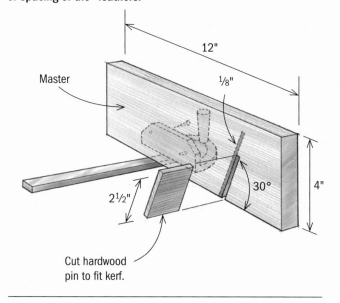

Master

12"

⅛"

30°

4"

2½"

Cut hardwood pin to fit kerf.

2. Clamp or screw your feather "master" to your miter fence. Raise the blade to 2½" and make the first cut. Rip a scrap of hardwood to the exact width of the cut you just made and glue it into the kerf. (The pin doesn't need to be exactly the height of the cut. It simply provides a registration point as you cut the individual fingers.)

3. Reposition the master so that the pin is roughly ⅛" away from the outside face of the blade and make a second cut, as shown in photo B. Realize that the thickness of the fingers affects the flexibility. For a stiffer featherboard, increase the distance between the pin and first cut or lower the height of the blade.

4. Crosscut the end of your future featherboard to 30°. Next, clamp it against the pin and feed it past the blade. Turn off the saw and wait for the blade to stop. Reposition the board so that the first kerf rides on the pin. Keep cutting fingers, as shown in photo C, for as long as you like. Trim wide featherboards if you need something smaller in the future.

Miter-Slot Featherboards

A miter-slot featherboard is similar to the one you made earlier, but it is designed to attach to a hardwood bar so it can lock itself into the miter slot on your tablesaw or router table. To avoid splitting the bar or featherboard when it's tightened in the slot, make both pieces from any straight-grained hardwood.

Drill and counterbore the hole for the machine screw in the bottom bar then use your tablesaw to cut the groove for the featherboard. First, draw a stop line on your board, set your rip fence, and cut along the line. Turn off your saw when the blade is about ¼" from your mark (the blade is cutting farther along the bottom edge). Adjust your fence and make the second cut. Use a coping saw to finish the groove.

Clamps control the cut. Be sure that the clamp is safely out of the blade's path as you make the cuts.

Slot-Locking Featherboard

The expanding bar provides adequate clamping pressure to hold the jig on tablesaws, router tables, or any other shop-built jigs with a ¾"-wide slot.

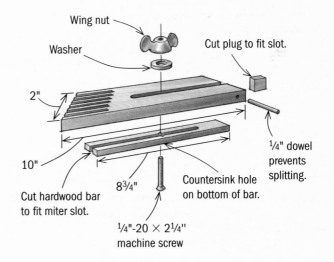

Wing nut

Washer

Cut plug to fit slot.

2"

10"

8¾"

Cut hardwood bar to fit miter slot.

¼"-20 × 2¼" machine screw

Countersink hole on bottom of bar.

¼" dowel prevents splitting.

Building the Legs

Gluing up thin boards is a perfectly acceptable way to make the thick stock needed for the legs and feet, but you want to orient the grain so that future wood movement doesn't force open the joint and create an unsightly gap. Wood tends to cup toward the bark side of the tree (if you need proof, take a look at any old deck). Arrange the lamination bark side to bark side so that the "inside" face of the tree faces out. (Check the end grain to determine which is the bark side.) That way, if the wood cups, the joint line along the outside edges will stay tight.

1. The 1½"-thick legs are made by laminating two pieces of ¾" stock face to face. Start by cutting four pieces of stock to 5½" × 32". To keep the boards from shifting as you clamp them together, tack two or three brad nails in the center of the lamination, then nip off the ends so just a small point protrudes from the wood, as in photo A. This small point will lock the mating piece into position and keep it from shifting as you tighten the clamps. *Note:* If you plan to rip the lamination to nibble out the mortise (see step 3), position the brads so they won't get hit by the sawblade.

2. Scrape off any dried glue squeeze-out along the edges of the stock, then make a light ripping cut at the tablesaw to establish a straight edge on one side.

3. There are two ways to make the mortise for the stretcher. If you would prefer to go the

Slip stoppers. Small brads prevent leg pieces from slipping. Drive the nail about ⅛" into the wood, then snip off the head to leave a sharp, short tip.

Mortising made easy. By ripping the leg into two pieces, you can treat the mortise like a dado. Clamp the board to your miter gauge and carefully saw up to your layout lines.

Smooth sliced tenons. For wide tenons, a tenoning jig out-performs nibbling away with a dado cutter. The jig cleanly cuts the cheek in one pass, and you won't need to waste time switching blades.

WORK SMART

You can sometimes use screws to apply extra clamping pressure or hold a board in place so you can move a clamp elsewhere. Drill pilot holes so that the screw's threads don't catch the top board; otherwise, the screw won't pull the boards together.

D

Cutting tapers on a plywood sled. **There are slicker tablesaw tapering jigs, but plywood works well enough for a few cuts.**

WORK SMART

When exposed to air and light, cherry's color deepens from a pinkish-tan to a deep red-brown. This change lends a desirable effect unless you have boards that are a few shades different because one was stacked on top of the other. A few weeks in bright sunlight should even out the color.

When done, reattach the cut strip. Once the glue cures, rip the leg to 4½". When you cut the leg to width, remember to center the mortise on the leg.

5. Once you've completed the mortises, cut the legs to 30". Remember that the bottom of the mortise needs to be positioned so that it begins 6½" up from the bottom end of the leg.

6. The next step is to lay out the taper and the tenons on each leg, as shown in photo C. Cut the tenons using your crosscut saw and tenoning jig. For more on this process, see "Skill Builder: Cutting Tenons on the Tablesaw" on pp. 268–269.

7. To taper the leg, use the simplest tapering jig of all—a ¼"-thick by 36"-long plywood sled, as shown in photo D. First, rip the plywood to 6". Without adjusting the fence, attach the leg to the sled with double-sided carpet tape so the cut line hangs over the edge of the plywood. (You can attach a few glue blocks with hot glue to secure the leg stock in place.) Adjust the blade height to cut all the way through the leg, then push the assembly past the blade. Use mineral spirits to release the tape and glue bonds. Repeat on the opposite side. Use a handplane or jointer to remove any saw marks.

traditional route, rip the leg to 4½" wide and turn to "Skill Builder: Cutting Through-Mortises" on pp. 264–265. Another option is to slice the leg along one side of the mortise, nibble it out with your tablesaw, then glue the pieces back together. The second method is easier, but sharp-eyed critics may see a glueline along the face of the leg.

4. To use the rip-and-nibble method, start by jointing one edge. Then, set your saw's fence to 2" and rip off one strip. On the wider piece, lay out the ½"-deep by 4"-long mortise (except now it's a dado) on one face of the leg. Raise your blade to ½" and carefully remove the wood between your lines, as shown in photo B.

SKILL BUILDER: Cutting Through-Mortises

What You'll Need

■ Combination square

■ Marking knife

■ Drill press and Forstner bits

■ Freshly sharpened chisels

■ Small piece of hardwood

■ Two 6" clamps

Traditionally, mortises are chopped out with just a chisel and mallet. While it's not particularly difficult, it can be slow going, especially if you're trying to make your way through hard or thick stock. Here, you'll learn how to use a drill press to tackle most of the work. After drilling out most of the waste, you can use a chisel to pare to the line. To cut down on some of the chiseling work—which gets old if you're cutting multiple mortises—you can use a router jig for hogging out waste.

1. Establish a centerline, then lay out the location of the mortise. Using a combination square, extend your lines from one face around the edge and onto

B

Calling good backup. Clamping a board against the outside edge of your mortise helps guide the chisel as you make the cut and prevents you from making the mortise too wide.

the opposite face of the board. Use a knife, rather than a pencil, when marking the edges of the mortise. The knife scores the surface of the wood, preventing tearout when drilling and provides a stop line that you can register against when chiseling.

2. Set up your drill press with the appropriately size bit. (For narrow mortises, you might use a brad point, but for the 1/2"-dia. and 3/4"-dia. mortises in this project, you'll use Forstner bits.) If your drill press doesn't have a fence, clamp a board to the table so the tip of the bit runs along the centerline of the mortise. Drill out the ends first and work toward the center, as shown in photo A. It's okay to overlap the drill holes, but position the bit so the center spur sits in solid wood.

3. Clamp a hardwood guide along the edge of the mortise, and guide the back of the chisel against it to remove the crescents left by the bit, as shown in photo B. To avoid tearing out the opposite face or edge,

A

Drilling does the hard work. A small benchtop drill press is fine for taking big bites out of mortises. Clamp a temporary fence in place so that you can concentrate on adjusting the board from side to side.

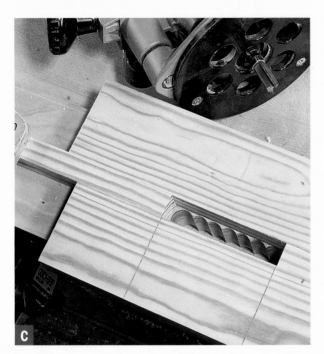

A solution for multiple mortises. **When you have more than a few mortises, consider using your router and this simple jig. The bit's bearing rides against the wood while the carbide cutters clean out the mortise underneath.**

or overchiseling your mortise, stop when you're halfway through the thickness of the material. Then flip the stock and finish paring out the waste from the opposite side.

Or Clean Up with a Router

With the jig shown in photo C, you can use your router and a pattern-cutting bit to do most of your cleanup work. The jig takes a few minutes to make, but it can save time if you're cleaning up multiple mortises. Rip a strip of wood to the thickness of the mortise and glue it between two pieces of wood. After drilling out most of the waste, clamp the template in place and feed the router in a counterclockwise direction. You'll still need a chisel to clean up the corners.

Building the Feet

Even if you could find 1½"-thick stock, it would still be easier to make the curved feet from a two-piece lamination. For starters, it would be difficult to chop a deep mortise in the top of the foot. Second, a solid foot would be apt to split along a weak grain line. A two-ply leg is less likely to crack.

You'll want to start with a full-size template. Using your template as a tracking guide will ensure that all four boards used to make up the feet are identical. Instead of sanding to the line, use your template to guide a router bit to remove saw marks and finish off the curves.

1. Cut four pieces of ¾" stock to 23¾". (That way, you can get all four pieces from one 8'-long board.) Joint one edge and rip each piece to 8½" wide. Next, use your tablesaw to cut a ⅜"-deep by 3½"-long dado along the center of each board.

2. Cut a spacer just a hair under ¾" thick and 3½" wide to help align the dadoes as you glue up the legs. Wrap the spacer with packing tape so that it doesn't become a permanent fixture, as shown in photo B on p. 267.

3. While you're waiting for the leg laminations to dry, use the dimensions given in the drawing on p. 266 to draw a full-size copy of half a foot on a piece of hardwood or MDF. Sand and/or file up to your pencil lines. This piece will serve as your template for shaping the feet. (The exact dimensions aren't critical. Just try to make the curves as smooth as possible.)

> ### WORK SMART
>
> **U**sing a template to transfer dimensions to both legs, instead of measuring both legs separately, can prevent measurement errors. Lay out the legs' dimensions, the taper, and the locations of the mortises and tenons on a piece of ¼"-thick hardboard or plywood and cut to the lines.

A

Mortising the feet. **After carefully cutting both sides of the dado (later to become a mortise), remove the wood between the cuts by making a few extra passes.**

4. Measure and mark the center of the mortise on the leg blanks, then position and trace the template. To help your bandsaw or jigsaw round the inside corners, drill relief holes in the tightest parts of the curves near the top and bottom of the foot. Try to cut about ¹⁄₁₆" to ⅛" along the outside of the line.

5. You can sand or file the saw marks off the foot boards, but if you have a bearing-driven straight bit with the bearing on the far end of the bit, often called a flush-trimming bit, you can use your leg template to finish shaping the foot at the router table (photo C). Attach the template to the foot with carpet tape and adjust the height of the bit so that the bearing rests against the template. Flip the template to shape the other side of the foot.

Foot Pattern

Use the given dimensions to establish the top and bottom edges. To establish a pleasing curve, connect those points with a flexible strip of wood.

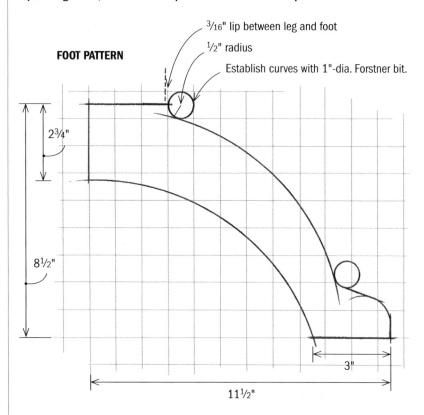

FOOT PATTERN

³⁄₁₆" lip between leg and foot

½" radius

Establish curves with 1"-dia. Forstner bit.

2¾"

8½"

3"

11½"

Foot Detail

The dadoes start as a guide when laminating the two pieces of the foot and later become a through-mortise once the glue dries.

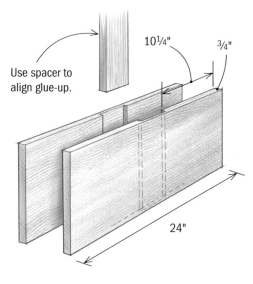

Use spacer to align glue-up.

10¼"

¾"

24"

Screws make good clamp companions. **Screws lack the clamping power of bar clamps, but they can hold two pieces tightly together so that you can move clamps where they're needed.**

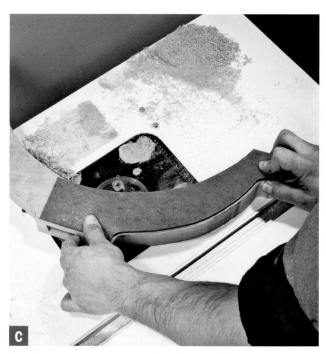

Double-duty template. **Made from ¼"-thick hardboard, the template used to trace the legs can also serve to guide a bearing-topped 1½"-long flush trimming bit. When routing, make only light cuts.**

Building the Stretcher and Top Supports

Use the dimensions given in the materials list and overall drawing as a starting point, but make a habit of transferring dimensions from pieces you've already cut onto the pieces you haven't. You may also want to make the tenons a hair longer than needed, then plane them flush after assembly. The stretcher's tenons should be slightly narrower than the mortises they'll fit into so the wedges have room to work.

1. To make the stretcher, rip a ¾" board to 5" and crosscut it to 27" long. Lay out and cut the ½"-wide by 4"-long tenons as you did with the legs.

2. Draw the location of the kerfs for the wedges ½" from each end. Drill a ⅛"-dia. hole to serve as a relief hole to prevent the wood from splitting, as shown in photo A (below), then cut the lines with a dovetail saw.

3. To make the wedges, start with a 6"-wide board and cut off a 3"-long piece. Reset your miter saw to make a 5° cut and position the board you just cut so the end grain (the wide edge) rests against the fence. Cut the angle on

Continued on p. 270

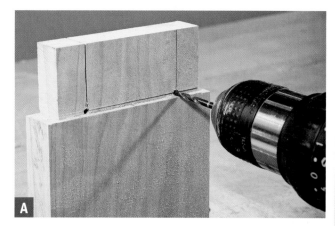

Split stoppers. **Drilling a ⅛"-diameter relief hole at the base of the tenon allows the wedge to flare the tenon open (creating a tighter joint) as it's driven in—instead of splitting the stretcher.**

SKILL BUILDER: Cutting Tenons on the Tablesaw

What You'll Need

- Crosscut sled
- Stop block
- Tenoning jig
- Tablesaw
- Small clamp

There are several ways to cut tenons, but this one is quick and easy to master. The five-piece tenoning jig shown in the drawing on the facing page can be built in a few minutes, but it works as well as store-bought fixtures costing a lot more.

Using this jig and your miter fence or crosscut sled, cutting tenons is a two-step operation. First, you'll establish the top, or shoulders, of the tenon. Next,

B

Cutting the cheeks. **The tenoning jig rides on your saw's rip fence so that the board is completely supported through the cut. For big boards, two clamps are better than one.**

you'll readjust the fence, raise the blade, and cut the sides (also called "cheeks"). You'll learn how to sneak up on a perfect fit using a plane or chisel.

1. Lay out the tenon on your stock and position it on your crosscut sled so the blade is correctly aligned. Clamp a stop block to the front fence to ensure that the shoulders on the second side (and on all subsequent tenons) are cut to the same height as the first, as shown in photo A. Next, set the height of the blade. To avoid dangling offcuts when cutting the cheeks in the next step, set the blade a hair higher than your layout lines. The overcut will be hidden and won't have a serious effect on the strength of the joint.

2. Place the tenoning jig over the rip fence and adjust the fence to cut along the outside of the line you made for the cheek cut. Adjust the blade height about $1/32$" below the shoulder line. Clamp the stock to the

A

Cutting the shoulders. **A crosscut sled is good for much more than its name suggests. Adding a stop enables you to use it for cutting tenon shoulders.**

Pared to perfection. Using a shoulder plane or chisel, shave just enough wood from the tenon's cheeks so that it slides into the mating mortise with a light tap.

jig, then push the work past the blade, as shown in photo B. Turn off your saw before attempting to lift or back up your work. Reposition the board and cut the opposite cheek.

3. A shoulder plane is the best tool for trimming tenons, as shown in photo C, but you can also use a wide chisel. Hold the chisel bevel up, as you would hold a knife. With your other hand, anchor your knuckle against the front edge of the workpiece and pinch the blade with your index finger and thumb. Using your index finger to support the blade and your thumb to control the depth of cut, shave off a thin layer of wood (see photo C inset), then test the fit against the mortise.

Tenon-Cutting Jig

This jig quickly and smoothly slices tenon cheeks in one pass. The fence prevents tearout on the back face of your work.

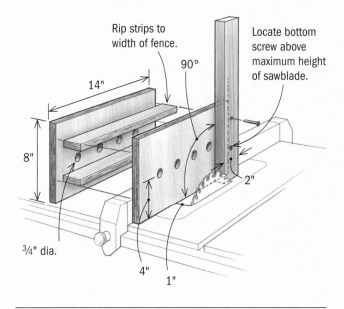

Rip strips to width of fence.

Locate bottom screw above maximum height of sawblade.

90°

14"

8"

2"

¾" dia.

4"

1"

Walnut wedges. By setting your miter angle 5° to the right and flipping the piece before each cut, you'll make perfect 10° tapered wedges. The wedge should have a fairly sharp point to fit into the kerf.

the edge of the board, then flip the board to make a wedge with a 10° angle. Make sure you clamp the stock in place, as shown in photo B above, because doing otherwise would mean holding your hands too close to the blade.

4. Cut two top support pieces to 2¾" by 18". Next, lay out a ¾"-wide by 3"-long mortise on the center of each board. (You can use the rip-and-nibble method as you did with the legs; however, because these mortises will be hidden by the tabletop, chopping these mortises by hand might be a good opportunity to work on your chiseling skills.) After cutting the mortises, use your router table to chamfer the bottom edges, then miter the ends at 45°.

5. The end holes of the supports are designed to allow the top to expand or contract in response to changes in humidity. Position the top supports bottom-face up, and drill four ⅛"-dia. pilot holes. Next, flip the piece and enlarge the end holes by drilling a ¼"-deep by ⁵⁄₁₆"-dia. hole. The "flared" bottom gives the screw room to move.

Assembling the Base

You're now ready to assemble the base. Take a moment to hone your favorite chisel—you'll probably have to take a few shavings off a tenon or two. Remember to dry-fit each sub-assembly. Finish-sand (up to 220 grit) any spots that might be difficult to reach once everything is assembled.

1. Apply a light coat of glue to the leg's bottom tenon and slide it into the mortise on the foot. Clamp the assembly so that there's no gap where the two pieces meet. From the inside face, drill two 1"-deep holes with a ¼" brad-point bit, as shown in photo A. Apply glue to short lengths of ¼" dowels and tap them into place. With that, your mortise-and-tenon joint has turned into what is called a pegged mortise-and-tenon joint. The pegs will keep the joint even more secure. Using a flush-cut saw, trim the dowels flush after the glue has dried.

2. At this point, a dry-fitting should show you legs that are perpendicular to the stretcher and parallel to each other. If not, adjust the tenon shoulders before glue-up. Once you're happy

Pegged into submission. Tenons were pegged into mortises long before glue came in neat plastic jugs. Simply drill a hole and tap in a length of dowel that has the same diameter. You can remove the clamps as soon as the dowels are tapped in place.

Wedged tight. Driving the wedges flares the end of the tenon, preventing the stretcher from coming loose. When the glue has dried, trim the tenon and wedges, then plane or sand flush with the leg.

with the fit, glue the stretcher into the legs. Use clamps to squeeze the legs together so that the stretcher's shoulders fit tight against the inside of the legs. Once clamped, wipe a small amount of glue onto the wedges and tap them in place, as shown in photo B. Trim the ends of the tenons and wedges flush.

3. Place the leg assembly on a flat surface—like your workbench—and place the top supports in place. Use a 2' level to make sure the supports are level before gluing them in place.

> **WORK SMART**
>
> **B**ecause the teeth are set on one side of the blade, a flush-cut saw is best for trimming protruding tenons and wedges. If you don't have one, slip a cardboard spacer between the blade and the wood.

Building the Bottom Tray

The bottom tray is what makes this project different from any run-of-the-mill table. Besides offering an extra storage shelf, this subassembly gives you the opportunity to show off your woodworking skills. The single dovetailed corner is sure to attract a great deal of attention. And, as you'll see, cutting a dovetail isn't nearly as tricky as it might appear.

You'll need to make a few cuts by hand, but you'll do most of the work with your tablesaw. You can adapt the techniques shown here to cut multiple tails, such as for a drawer or blanket chest. For wider panels, consider building a designated crosscut sled just for cutting the angled tails.

1. Start by ripping a board to 3½" to make the ends and sides of the tray. Crosscut the sides to 20" and the ends to 16". Cut a ¼"-wide by ½"-deep groove along the lower edge on the inside face of the side pieces, as shown in the drawing on p. 272. Cut the boards for the bottom of the tray to 3¼" wide by 15½" long. (You may also substitute a solid piece of ¼"-thick plywood for the bottom.)

2. Using a marking gauge, find the thickness of the sides and ends and add a hair (less than ¹⁄₁₆") to that measurement. Run the gauge around the ends of all four boards. You'll use this baseline to establish the cutting height of the pins and tails.

3. Transfer the dovetail to a piece of cardboard or aluminum flashing to make a longer-lasting guide. On the sides of the tray, line up the shoulder of the template with the baseline you marked with the marking gauge. Then trace tails on both ends, as shown in photo A on p. 272.

Tray Detail

Single-dovetailed corners spice up an otherwise simple box. This mini-project is a good opportunity to practice this classic woodworking joint.

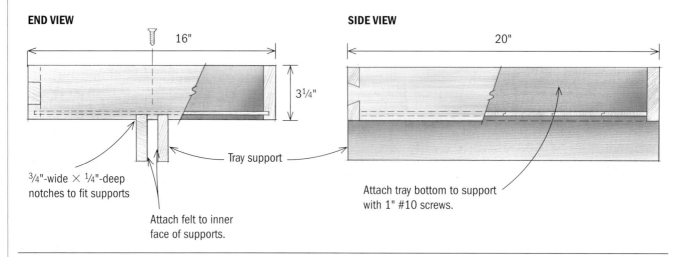

END VIEW

16"

3¼"

Tray support

¾"-wide × ¼"-deep notches to fit supports

Attach felt to inner face of supports.

SIDE VIEW

20"

Attach tray bottom to support with 1" #10 screws.

A

A tin tail template. Making a template from a scrap of flashing is faster than measuring out the tail on four separate ends. Note how the tail misses the groove for the bottom panel.

4. Cut the tails on your tablesaw. Clamp the sides to your tenoning jig, then adjust the bevel of the blade until it matches the line. Set the blade height just shy of the shoulder line and make the cut, as shown in photo B on the facing page. After cutting both sides, finish the shoulders with a dovetail saw.

5. Clamp the end board in a vise so that the end of the board is slightly above the benchtop, and set the tail board on top. To mark the pins, transfer the location of the tail onto the end grain of the end piece, as shown in photo C. Use a combination square to extend the pencil marks you made on the ends down both faces to the baseline.

6. Use a dovetail saw to cut along the waste side of the line, down to the baseline, as shown in photo D. After cutting both sides, remove most of the waste up to the baseline with a coping saw (see photo D, inset) and chisel. Test-fit each corner and shave off any high spots that prevent the tail from fitting between the pins. When test-assembling, pull the pieces straight apart; levering them will affect the appearance of the joint and can snap a pin or tail.

7. Assemble one side of the tray, then slide the bottom panels into the groove. If you made a rabbetted bottom, you may need to trim the panel pieces so they fit between the ends. If that's the case, temporarily remove all the bottom panels, then trim half the excess from the

Sawn to perfection. Here your tenoning jig does double duty. Adjust the angle of the blade to match the tail, and make the cut. Cut just below the baseline and finish with a handsaw for a tight-fitting joint.

Tracing ensures a tight fit. By using the tails instead of a template, there's no need to worry if your cut is slightly off your original lines. Trace each tail's mating piece.

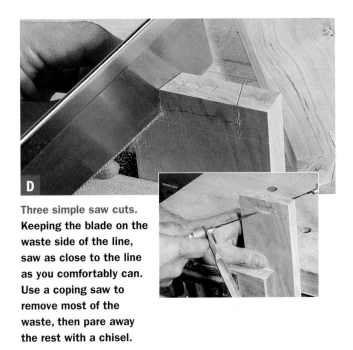

Three simple saw cuts. Keeping the blade on the waste side of the line, saw as close to the line as you comfortably can. Use a coping saw to remove most of the waste, then pare away the rest with a chisel.

Plane away the evidence. Skewing the plane—so that the blade hits the end grain at an angle rather than straight on—helps slice through difficult end grain.

Custom-fit tray supports. To lay out the notches, clamp the tray supports on either side of the lower stretcher and balance the finished tray on top.

outside edges of the two outermost panels. Doing this keeps the grooves between the boards even.

8. The tails and pins were deliberately cut a little long to make it easier to clean up the joints. The quickest way to do this is with a freshly honed handplane. As you plane across the end grain, work from the edges toward the center, as shown in photo E. If you attempt to plane straight across, you may cause the outside unsupported edge to splinter.

9. Cut the tray supports to size and clamp them to either side of the stretcher, as shown in photo F. Center the tray on the stretcher and mark the point where the supports intersect the ends. Use a chisel to cut the four ¼"-deep notches so that the tops of the supports are flush with the bottom of the tray. Attach the tray to the supports with glue and eight ¾"-long screws driven through the bottom of the tray.

Finishing

Here's how to apply and rub out a varnish top coat to achieve a nice satin sheen. With additional sanding using higher-grit papers and polishing compounds, you can get a high-gloss finish. Just realize that higher sheen will cause even the smallest imperfection to stand out.

Because you intend to abrade the surface, you'll want to use a hard-film-forming varnish. Technically, polyurethanes and oil/varnish wipe-ons are part of the varnish fam-

ily, but these finishes are too flexible or don't leave enough of a surface film to be rubbed out. Names that include the words *rubbing* or *rock hard* are good clues, but to be sure you're using the right stuff, look for rubbing-out instructions on the back of the can.

1. Using a random-orbit sander, sand all surfaces up to 220 grit. Use a tack cloth to remove any dust from the project. Wipe down your bench. It's also a good idea to wet-mop your shop floor and put on a fresh work shirt to get rid of dust particles that would otherwise stick to your finish.

2. Mix a batch of varnish 50:50 with thinner to improve the penetration of the first coat. Apply the varnish across the grain, then lightly brush with the grain. After the entire surface has been coated, tip off the varnish by lightly dragging

Even scratches make a satin finish. Using 0000 steel wool wrapped around a wood block, rub the perimeter first, then do the center section using long, straight strokes. Try to keep the pressure uniform and your strokes parallel to the grain.

the bristle tips through the wet finish. Don't spend too much time trying to level out brush marks; minor ripples should level themselves out or will be sanded down in later coats.

3. Let the first coat dry for at least 24 hours, then sand it using a random-orbit sander outfitted with 320-grit sandpaper. Try not to sand through the first coat; the goal is to knock off bumps and provide some tooth for the next coat. If necessary, use a sanding block to flatten thick drips. Wipe down the surface with a tack cloth before applying the next coat.

4. Brush on a second coat like the first—using long, smooth strokes across the grain to lay on the varnish, then smoothing it out with long strokes along the grain. Continue the varnish and sanding routine for at least three coats. (If you accidentally sand through one of the coats, plan on four or five.)

5. After giving the final coat a week (at least) to cure, knock off any bumps with 600-grit paper. Next, wrap a piece of 0000 steel wool around a sanding block. Rubbing with the grain, make three or four complete passes over the surface, slightly overlapping each pass with the next. Apply a small amount of lemon oil or soapy water to the board to lubricate the steel wool as you give the board a final rubdown, as shown in the photo on the facing page. When dry, buff off any remaining haze with a clean rag.

Final Assembly

To protect your top, place a blanket on your bench before laying the top face down. Position the leg assembly so that there's an even overhang along the edges and ends.

Using the previously drilled holes in the top supports as a guide, drill pilot holes in the underside of the top, then screw the base to the top. When driving the screws, back them off by a

Attaching the top. Use an old blanket to protect the tabletop as you install the final screws. Unless you're careful, dried glue or a few specks of abrasive can wreak havoc on your top in an instant.

quarter to half turn. While you want the top to be securely attached, you should allow for a little wood movement.

Face the inside of the tray supports with felt or thin cork. The lining will make the supports fit a little tighter and will prevent them from scratching the wood.

At this point, you can bring the table in from the shop and make it start working for you. Periodically, wipe on a light coat of wax. Wax provides little protection, but it helps keep stains from sticking. It will also help your project look its very best.

Simple Bed

When you shop for a bed, you'll discover that commercial furniture makers are demanding premium prices for handcrafted details such as "eased corners," "hand-sanded surfaces," and "oil-and-wax finishes"—stuff you can do equally well in your shop. Isn't it time that you started making your own heirlooms instead of buying them at the mall?

A big project like a bed can be boiled down into just a few simple parts: a headboard, a footboard, and a pair of rails to tie the two together. Once you understand the basic construction process, even making changes to fit your personal style isn't a big deal. You'll also learn ways to shrink the footboard, remove the slats, and even create a frame-and-panel footboard.

This bed was designed to fit into almost any bedroom. One way to do this is to take design cues from a few different styles. The rectilinear slats and legs are in keeping with solid Arts and Crafts furniture, but the simple form, and the use of cherry instead of oak, evokes a lighter Shaker feel. Another way to make a project fit is by designing it to complement existing decor without competing for attention. This headboard and footboard were made slightly smaller than many Arts and Crafts beds so that they wouldn't overpower other pieces in a small room.

This project was built to accommodate both a mattress and a box spring, but these days you can get by with just a mattress. In that case, consider raising the ledger strips so that the support planks sit just below the top edge of the side rails.

What You'll Learn

- Handling big boards in a small shop
- Using epoxy to hide blemishes
- No-mortise mortise-and-tenon joinery
- Cutting wide dadoes with a router
- Using planes and chisels to make a finished surface

While beginning woodworkers will have an enjoyable time tackling this project, it offers plenty of opportunities to learn a few new things regardless of your level of expertise.

If you haven't been introduced to no-mortise bed-rail fasteners, you're in for a pleasant surprise. Instead of drilling holes through the legs for a bed bolt or making a mortise for a metal hook, these fasteners simply screw to the legs and rails and rely on gravity to lock them together. These inexpensive fasteners are solid, yet easy to knock apart when it's time to move the bed.

You'll learn a technique for making the multiple mortises for the slats in the headboard and footboard without endless drilling or chiseling. Instead, you'll build a router jig designed to rout wide dadoes across one board, then cut the notched board into strips that fit into headboard and footboard rails. Turning the dadoes into a ready-made mortise strip ensures tight-fitting slats and perfect top-to-bottom alignment.

Finally, you'll have an opportunity to fine-tune your plane and chisel skills as you finish the legs and rails. The chisel-chamfered legs and hand-planed surfaces may not appear as "perfect" as a piece that's pushed through a stroke sander, but when done well, the surface will gleam under a coat of oil. Once you've run

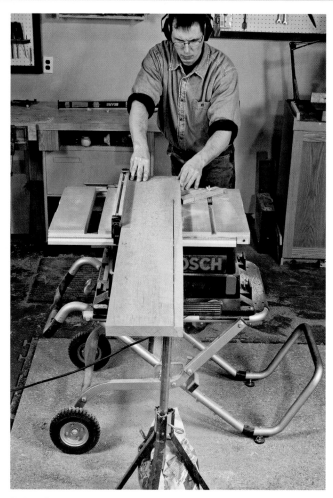

Balancing acts with benchtop tools. Learn how to safely manage long boards on small saws by providing support before, during, and after the cut.

Make your own mark. Hand-cut details, like chamfered edges, add visual and tactile interest that you can't get from power tools.

A Flexible Bed Design

This bed is fine as is, but both the dimensions and the construction methods can be used as a template for making several different beds. Metal hardware and mortise strips simplify the trickiest construction steps.

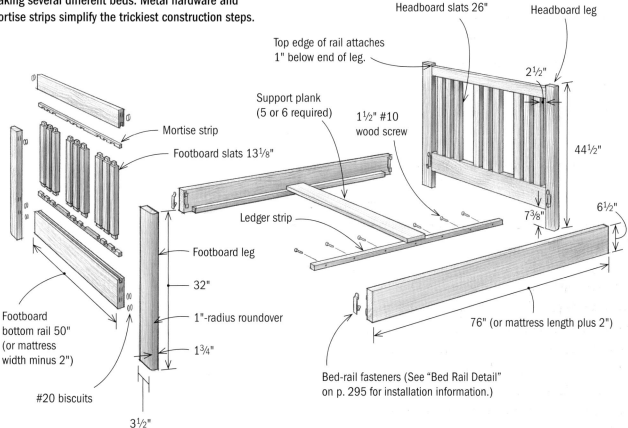

Headboard slats 26"

Headboard leg

Top edge of rail attaches 1" below end of leg.

Support plank (5 or 6 required)

1½" #10 wood screw

2½"

44½"

Mortise strip

Footboard slats 13⅛"

Ledger strip

Footboard leg

32"

1"-radius roundover

1¾"

7⅜"

6½"

Footboard bottom rail 50" (or mattress width minus 2")

#20 biscuits

3½"

76" (or mattress length plus 2")

Bed-rail fasteners (See "Bed Rail Detail" on p. 295 for installation information.)

your fingers over the tiny facets left on a hand-tooled surface, you may find that there's no going back.

Buying Materials

Building a bed requires fairly big boards. The legs and rails are made from 8/4 and 6/4 stock, respectively. If the milled wood is slightly thicker than what's listed in the materials list, stop while you're ahead—the added thickness will add only a little extra visual weight to your project. Just make sure that the boards used for the legs are all one thickness, and the material for the rails is about ¼" thinner.

The mortises are cut as a strip from a piece of ½"-thick wood. If you don't yet own a planer, ask your mill operator to plane down an 8' board for you. (Ideally, the finished board should be about 9/16" thick; you'll plane it to exact thickness once it's glued in place.)

Assembling the headboard and footboard will require a pair of long clamps. One pair of 60" K-body clamps is a good investment, but if you already have pipe clamps, you can save money by buying an appropriately long piece of black pipe.

MATERIALS

Quantity	Part	Actual Size	Notes
2	Side rails	1¼" × 6½" × 76"	6/4 hardwood Don't worry if the two rails don't match perfectly; few people will ever see both at once.
2	Ledger strips	¾" × 1" × 73"	Use whatever hardwood stock you have on hand.
2	Headboard and footboard bottom rails	1¼" × 6½" × 50"	6/4 hardwood Cherry, walnut, oak, or maple would all work well.
2	Headboard and footboard top rails	1¼" × 4½" × 50"	6/4 hardwood Position the top and bottom rail stock next to each other (before cutting to length) to ensure a pleasing grain match.
2	Headboard legs	1¾" × 3½" × 44½"	8/4 hardwood You probably want the legs to match the rails.
2	Footboard legs	1¾" × 3½" × 32"	8/4 hardwood
9	Headboard slats	¾" × 2¼" × 26"	Cherry was used here, but you can use a different species for a nice contrast. Cut a few extra to ensure good grain match from slat to slat.
9	Footboard slats	¾" × 2¼" × 13⅛"	Cut a few extra to ensure good grain match from slat to slat.
4	Mortise strips	9/16" × ½" × 50"	Cut strips 2" long and trim to fit after gluing into rails.
5 or 6	Support planks	¾" × 5½" × 52⅞"	Pine or poplar is fine. Space them evenly across the ledger strips to support the box spring.
24	#20 biscuits		Buy a container.
14	1½" #10 wood screws		To attach ledger strips to side rails
4	Bed-rail fasteners		See Resources on p. 296.
32	1" #8 wood screws		To attach bed hardware to side rails and legs
	Yellow glue		Extended or slower-setting glue can be handy for assembling headboards and footboards.
1 pint	Oil/varnish finish		The tops of the legs will wick in oil like a sponge.
	Miscellaneous		Wax, blue masking (painter's) tape, white synthetic wool pad, cotton rags, 150- and 220-grit sandpaper (sheets and disks)

MATERIALS – SHORT-STYLE FOOTBOARD

Quantity	Part	Actual Size	Notes
2	Short legs	1¾" × 3½" × 15"	Two 8/4 short legs can be made with less material than it takes to make one headboard leg.
1	Footboard rail	1¼" × 6½" × 50"	
8	#20 biscuits		

TOOLS

- Tape measure
- Tablesaw with rip blade and dado cutter
- Miter saw
- Biscuit joiner
- Drill press (optional)
- 12" combination square
- Block and #4 or #5 handplane
- ½" or ¾" chisel
- Card scraper
- Planer (optional)
- Router table
- 1"-radius roundover bit
- ½"-dia. straight bit
- Bearing-guided chamfer bit
- ⅛"-dia. drill bit
- ⅜"-dia. countersink bit
- 7/32"-dia. centering drill bit
- Drill
- Screwdriver
- Marking knife
- Marking gauge
- Two 60"-long bar clamps or two strap clamps
- Several small bar clamps
- Random-orbit sander

Building the Bed

As you cut the parts, "read" the wood and use it to its best advantage. For example, if a board has a minor bow, it can't be used for the headboard or footboard, but it might work well on the side rails. Arrange leg, rail, and slat boards so that the grain works in a way that's complementary, not conflicting.

Take time to apply the finish before assembly. If the surface will be glued later on, apply a light coat of oil/varnish finish after you've planed or sanded it smooth. The finish will help protect the surface of the wood from fingerprints and glue drips and will point out any rough spots that you might have missed.

Start and Finish the Rails

As with the other projects, you'll cut the longest boards first. If you make a mistake, you'll still have a few big boards sitting around, and you may be able to salvage shorter pieces from your mistake.

1. Using the measurements from your mattress, cut the rails for the sides, headboard, and footboard to rough length. Normally, an extra 1" is more than enough, but in this case, you'll want to cut one bottom and one top rail 4" to 6" longer than needed. (You'll use the cutoffs later as slot-spacing gauges when cutting the biscuit slots.)

2. Joint one edge, then set your tablesaw fence to 6½" and rip the side rails and the bottom rails for the headboard and footboard. (For tips on cutting long boards on benchtop tools, see

> **WORK SMART**
>
> If your mattress is larger than standard dimensions, you won't be able to squeeze it into an undersize frame. Measure your mattress before cutting any stock and adjust your plans as necessary to fit your mattress. The frame's interior dimension should be 2" longer and 1" wider than the box spring or mattress.

"Skill Builder: Cutting Big Boards in a Small Shop" on p. 287.) Reset your fence to 4½" and rip the boards for the top rails. Place the headboard and footboard pieces safely aside for now.

3. Using your miter saw, or circular saw and edge guide, cut the side rails to length. The extra 2" allows some wiggle room for the mattress, so it's okay if you're off by as much as ¼". Just make sure that the rails are the same length.

4. Cut the ledger strips so they are 3" shorter than your side rails. After cutting them to size, drill, countersink, and attach the cleats to the rails using 1½"-long screws, spaced about 12" apart. (Because I own a box spring, I attached the strips so the bottom edge is flush with the bottom edge of the side rail. If you use only a mattress, consider raising the strips so the top edge is 1" below the top edge of the side rail.)

5. Knock the sharp edges off the rails and ledger strips. The fastest way to do this is with a router and bearing-guided chamfer bit, as shown in photo A on p. 282, but a block plane will work just as well.

Using what you've learned from the earlier chapters of this book, you can adapt this plan to make several different styles of beds. Dimensions and photos are given for both tall and short footboards, but for a different look, you can also omit the slats in the headboard and/or footboard.

If you prefer the frame-and-panel style as in the drawing shown at right below, groove the legs and end rails. Dry-assemble the frame to find the measurement of the center panel.

SLATLESS STYLE

Omitting the mortise strips and slats makes the bed look lighter and makes the project easier to build. You could build two twin-size frames in a long weekend to outfit a kids' room.

SOLID-PANEL STYLE

If you choose solid wood, widen the grooves in the legs and fasten the panel only in the center so that it's free to move across its width. Plywood panels eliminate the need to worry about wood movement.

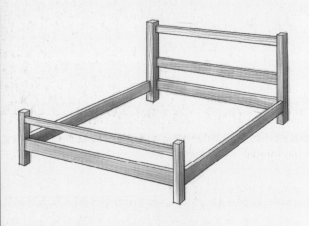

Better-balanced router. Unless chamfered, cherry and maple edges are sharp enough to slice skin. Balance the base of the router on scrap, as shown, when chamfering the outside edges.

6. Lay both rails on your bench so the ledger strips are facing up. Refer to the drawing on p. 295 to see how the hardware should be oriented and lay pieces onto the ends so that you have a left and right side. Use a centering bit to drill pilot holes, then screw the metal clips in place, as shown in photo B.

7. Before putting the rails aside, apply at least one coat of finish to protect them from glue and miscellaneous shop stains. Temporarily remove the hardware, plane and/or sand the rails up to 220 grit, then wipe on a coat of oil. Wait 20 minutes, then wipe away any excess. Apply at least three coats before topcoating with wax.

Quick clips. To ensure a gap-free connection, butt the side rails against a scrap of wood when attaching the bed-rail fasteners. That way, the metal edges can't extend past the end as you drive the screws.

Making the Legs

From a distance, the legs may appear massive, but they're only 1¾" thick—about the same thickness as standard 2-by lumber. Thick, square posts would add a more substantial look, but they won't make the bed any sturdier. Another consideration is price. The board foot price of 16/4 stock is usually more than twice the cost of 8/4 wood and isn't always available in all species.

To soften the outside corners, I used a roundover bit with a 1" radius. Large-diameter bits should be used only with larger (2 hp or more) routers that can be slowed down to a safe cutting speed. If you don't feel like springing for a big bit or don't yet own a router with speed control, you can knock off corners with a smaller bit or cut the bevel with your tablesaw, as shown in photo B on p. 284,

as shown in photo B on p. 284,

> ## WORK SMART
>
> Inspect both edges and ends of each board before cutting to final dimension. This is your chance to cut off natural defects or tearout, blade burns, and snipe marks left by your planer or other machines.

then use your block plane to finish the round-over process.

1. Start by cutting the legs to rough length (cut the headboard legs to about 46" and the footboard legs to about 33"). Joint one edge, then set your tablesaw to 3½" and rip all four pieces to width. Now that they're a more manageable size, trim the headboard legs to 44½" and the footboard legs to 32".

A

Laying out the legs. **Position the legs on your bench, inside face up, and mark the locations for the top and bottom rails. Use a scrap piece of rail stock to lay out the rail's location on the legs' inside edges.**

B

Precut corners. **Instead of overworking your router, or wasting time repositioning the bit or fence to make multiple cuts, set your tablesaw to 45° and slice off the outside corner before routing.**

2. Take a minute to decide how the legs will be oriented in relation to the bed. When you're happy with the way things look, use a pencil to mark the top end and the outside corner that you will round over. Next, position the legs on your bench so that the bottoms are against a straightedge and the inside edges face up, as shown in photo A. Using a combination square, lay out the locations of the bottom and top rails.

3. To produce a smoother corner and save your bit from needless wear and tear, use your tablesaw to remove some material from the corner first, as shown in photo B. Before you cut the legs, test-cut and rout a piece of scrap to be sure you're not sawing off too much wood.

C

Chew slowly and take small bites. **Don't think big bits can remove all that wood at once. To prevent damage to the bit or board, take light cuts and slow down the router to a safer, slower speed.**

4. Using your router table and 1" radius bit, round over the outside edge of each leg, as shown in photo C. Lower the operating speed of your router to prevent damaging you, your router, or your workpiece. (See "Safe Speeds for Big Bits" on p. 288.)

5. After softening the outside corner, knock down the remaining three sharp edges with a chamfer bit set just below ⅛" high, as shown in photo D.

6. Now for the fun part. Clamp one leg to your bench. Using one or more of your favorite planes, knock down any bump that might exist at the edge of the rounded-over corner, as shown in photo E on p. 286. Concentrate on knocking off any ridges along the corner to make the fairest curve possible. If your planes are working to your satisfaction, continue smoothing the out-

D

Chamfer corners before they cut you. **Armed with a bearing-guided bit, this simple router table is a champ at cutting light chamfers. Use it to soften the sharp edges from the legs, rails, and slats.**

side faces of the leg. Alternately, you can hand-sand or use a random-orbit sander to smooth the legs up to 220 grit.

Plane around the corner. **Set your blade to take light cuts, and focus on knocking off the ridge between the roundover and the flat face. The rounded edge will look smooth, but your fingertips will feel the facets on the legs.**

Chiseled chamfers. **Chiseling the chamfers on the ends not only eliminates the chance of burning or tearout, it also adds another interesting hand-tooled detail.**

7. You no longer need the pencil marks on the tops of the legs to know which end goes up. Sand, plane, or slice off the reference marks you made earlier (no one will know if your leg is $^1/_{32}$" shorter than the given dimension). Next, secure the leg to your bench, bottom end up. At this point, you're ready to chamfer the ends. The chamfer on the bottom end prevents the wood from catching and splintering; on the top, it's purely for decoration.

Cutting a chamfer with a chisel is easier than you might think. Start by making short, light cuts. Resist the chisel's inclination to drive itself too deeply into the wood. Aim for about a $^1/_8$"-high chamfer, as shown in photo F, but feel free to stop when it looks good. After chamfering the bottoms of all four legs, sharpen your chisel and do the top ends. (Alternately, you can make the chamfer using a file or sandpaper wrapped around a block of wood.)

Making the Headboard and Footboard

Headboard and footboard rails usually rely on tenons to attach them to the legs. Tenons are a good choice; they're solid enough to help keep the frame square and can be designed to allow single-plank headboards room to move in response to changes in humidity. Sometimes, however, you can achieve the same effect with a simpler approach. In this case, biscuits make a perfectly serviceable joint; the narrow rails can be glued tight to the legs because they don't move as much as a wide panel headboard. Because the rails transfer the weight of the mattress directly to the leg, strength isn't a concern.

SKILL BUILDER: Cutting Big Boards in a Small Shop

What You'll Need

■ Benchtop tablesaw

■ Miter saw

■ Roller stand (weighted down with sand)

■ Bench-mounted roller supports

One reason small-shop woodworkers stick with small projects is that they're uncomfortable wrestling big boards in a small space. Benchtop tools have the horsepower to make most cuts, but lacking the mass and footprint of larger machines, they may not be able to safely support the workpiece. Boards that topple off tools in midcut can result in damage to the work, the machine, or both. But if you plan out your cuts and provide support at both ends of each board, your wood won't know whether you're working in an aircraft hangar or a one-car garage.

Safe, Supported Rips

When ripping long boards on a benchtop tablesaw, a roller stand is a necessity for making smooth, safe cuts. Set the stand height so that it's about 1/8" below the saw table to prevent the board from catching the front edge. Position the stand close enough to your saw so it provides support before the board begins to tip off your saw but far enough so the board won't tip off the stand at the end of the cut. To provide extra stability, weigh down the stand's legs with a bag of sand, as shown in photo A.

Controlled Crosscuts

A miter saw can't cut accurately if the board tips or slips midcut. The saw's hold-downs help, but for really long boards, an auxiliary table or fence is a better bet. The supports shown in photo B can be clamped onto any sawhorse or tabletop. Alternately, you can build a riser box that can sit on your bench, to support long boards.

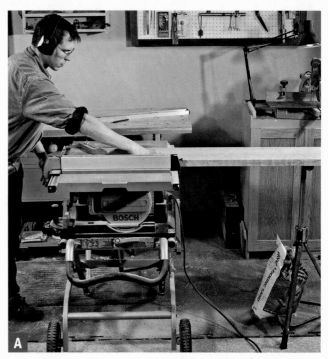

A

Better balanced board. Position the stand far enough away from the saw so that the board doesn't tip at the end of the cut. The sandbag ensures that the stand doesn't tip in midcut.

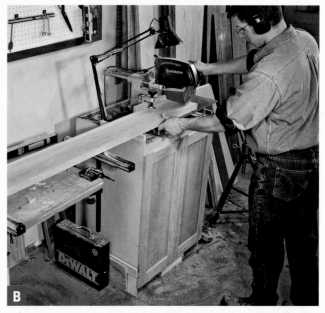

B

Double-duty workbench. Quick-clamping roller supports grab onto your bench or a sawhorse to support oversize stock. Use a 4' level to set the supports to the same level as the saw.

The speed at which a router bit turns can be important. The typical router runs at about 22,000 rpm, which is fine for most bits. But as the diameter of the bit increases, you'll want to slow things down. In the worst-case scenario, a big bit at full speed could spin itself apart. Before that happened, the bit would begin to flutter, which would cause even a brand-new bit to leave burned edges or a rippled finish.

When using big bits, decreasing the router speed can make routing safer and improve the quality of the cut. To the right is a chart showing recommended speeds for various sizes of bits. If you find yourself burning edges with fresh bits, you may choose to use a speed slower than what's listed below. To be extra safe, big bits should be used only in a router table.

Bit Diameter	Recommended rpm
1" or less	Full speed
1" to 1½"	18,000
1½" to 2½"	15,000
More than 2½"	10,000

Ready the rails for mortise strips. The mortise strips fit into a ½"-wide by ½"-deep groove. Use an auxiliary fence and featherboard at the tablesaw to prevent the rail from tipping in midcut.

Starting the rails and finishing the slats

1. Pull out the headboard and footboard rails you made at the beginning of this project and cut all four boards 2" shorter than the width of your mattress or box spring (the width of the legs makes up for the 2" that the rails come up short). Next, lay them out on your bench and pick out the best faces of each and arrange them to make the most attractive pairs. (You may want to mark the rails to make sure they don't get switched around.) Knots or other blemishes can usually be hidden simply by flipping the board, but if there's a blemish that you can't work or cut around, refer to "Skill Builder: Hiding Knots and Blemishes" on p. 292 to learn how to make a near-invisible patch.

2. Temporarily put the rails aside and start cutting the slats. To make the slats, begin by cutting the ¾"-thick wood to rough length. (Rather than cutting the boards to two different lengths at this point, cut all the slat stock 27" long. From that length, you can make one headboard slat or two footboard slats. Cut two or three more than you need, just in case.) Joint

Cut the slats to size. By clamping a long-arm stop to your saw, you can be sure that all your slats are the same length. Cut a few extra before adjusting or removing the stop.

Slip-fit stub tenons. Starting with a test slat, use your tablesaw and dado cutter to nibble away at both sides of the tenon until it fits the groove. Adjust the cutter height so that the tenon doesn't wiggle from side to side.

one edge, then set your tablesaw to 2¼" and rip the strips to final width.

3. Using your router and chamfer bit set to ⅛" high, or just your block plane, chamfer the long edges of the four rails and your slat stock.

4. It's easier to sand and finish the slats now, rather than attempt to work between them once they're joined to the rails. Finish-sand all surfaces up to 220 grit, then apply one or two coats of oil.

5. While you're waiting for the oil to dry, focus your attention on the rails. Replace your sawblade with the inner and outer blades from your dado cutter and set the cutter height to ½". Cut a ½"-wide groove centered along the bottom edge of the top rail and the top edge of the bottom rail. Start by making one pass, then turning the board end for end and cutting a second pass, as shown in photo C. (Don't lose any sleep if the groove isn't exactly ½" wide; because the mortise strips are cut to fit, you can still get a tight-fitting joint.)

6. When cutting the long and short slats to length, take a minute to set up a stop block system, like the setup shown in photo D on p. 289, to ensure that your slats are cut to the same length. Before moving or removing the jig, cut two or three extra long and short slats, just in case you make a mistake later on in the assembly process.

7. Adjust the dado cutter so that it's just ⅛" high and ¼" away from the fence. Next, using your miter gauge to guide a sample slat, cut a tenon on one end, as shown in photo E on p. 289. If necessary, you can adjust the fit by changing the cutter height. Once you've gotten a good slip fit, as shown in photo F on p. 289, don't touch your saw's settings until you've finished tenoning all your slats.

Making mortise strips

The mortise strip starts out as a 4½" by 52" board dadoed across its top face; it's then ripped into strips that fit into the grooves cut into the headboard and footboard rails. Making four rows of "mortises" at once is faster than cutting them separately, and it ensures the holes in the top and bottom rails line up.

1. Start by cutting the mortise board to rough size. Plane or resaw a board to ⁹⁄₁₆". Cut the board to 52" long (2" longer than your headboard and footboard rails). Draw lines at each end of the board to indicate the final length.

2. Use a 2½"-wide scrap spacer and one of the actual slats to lay out the dadoes. Center the first dado at the midpoint between your two pencil lines. Using the spacer board as a guide, start the second and third dado 2½" away from the center dado (on either side of the center dado). After laying out the three center slats, use the spacer to position the first end slat 2½" in from the pencil line and a slat to obtain the

Numberless layout. Use a 2½" spacer strip and your test slat to lay out the location of the dadoes on your mortise strip. Use your square to extend the lines, not to measure.

Mortises made easy. Make the "mortises" for all four rails at once by using your router and the wide dado jig.

exact width of the dado. Draw X's on the dadoes, as shown in photo G, to ensure that you don't accidentally rout out "good" wood.

3. After marking out the board, the next step is to cut the dadoes with either the tablesaw or the router. If you decide to try the router, as shown in photo H, use the jig described in "Skill Builder: Routing Wide Dadoes" on p. 294.

Planed to perfection. Don't try to get a perfect-fitting strip straight off the tablesaw. Rip the strip slightly wider than you need, then plane to fit.

Shave away the evidence. To remove remaining blade marks and glue stains, plane the strip flush with the rail. Don't get carried away; stop planing as soon as your blade starts shaving the rail.

4. To turn those dadoes into mortises, adjust the bevel angle of your tablesaw to 3°, then set your tablesaw's fence to rip a strip a hair thicker than the groove you made in the rail. When you rip the dadoed board, orient it so the notched side is wider than the unnotched side.

5. In theory, the strip should fit into its groove like a cork in a bottle, but it's okay if you don't have a perfect fit yet. If it's too tight—either across the entire strip or in just a few spots— plane down one edge, as shown in photo I, until the strip fits with just a light mallet tap. Make sure to align the end marks on the notched strip. Use as little glue as possible; you don't want squeeze-out to fill your brand-new mortises.

6. The mortise strip was made 1/16" thicker than the depth of the groove. Use a handplane to trim the strip flush with the edge of the rail, as shown in photo J. Set your plane for a light cut and balance it on the exposed strip. Stop planing as soon as the blade bites into the rail.

Assembling the headboard and footboard

1. Before you can start pulling the headboard and footboard together, you'll need to cut biscuit slots in the legs and end rails. Start by cutting a pair of biscuit slots in a scrap of 6½"-wide rail to see how far you can space the slots without coming out through the edges. Then transfer the slot locations to the leg stock, as shown in photo K on p. 293. Continue laying out the slot locations on the bottoms of all four legs and both ends of the bottom ends of both end rails. Repeat this step using the 4½"-wide rail scrap to position the slots for the top rail.

2. The rails are offset from the legs by ¼". To offset the biscuit slots, place a leg flat-side down against your bench, lay one ¼" spacer under your biscuit joiner, and cut the bottom slot. Insert two more ¼" spacers under the joiner and cut the higher slot, as shown in photo L on p. 293. Repeat on all four legs.

3. To cut the slots in the rails, place the rail against your bench, inside face down, and cut the biscuit slots, first with no spacer, then with two, as shown in photo M on p. 293. When the rails are fit against the legs, their biscuit slots will be offset ¼" in from the edge of the leg. Repeat the biscuit-cutting process on all four rails.

Continued on p. 293

What You'll Need

■ Plastic cup and knife

■ 5-minute epoxy

■ Sawdust

■ Chisel or card scraper

Cracks, checks, and knot holes are par for the course with wood. It's usually best to work around defects and dents, but when that's not an option, try using this epoxy and sawdust filler. Using sawdust from the wood you're repairing guarantees a close color match. Epoxy can be machined like the surrounding wood, and it works under most finishes. Even under a clear finish, the patch resembles a hard knot.

1. Follow the instructions on the epoxy to mix the resin and hardener. As you mix, blend in some fine sawdust (the type you'd shake out of your sander after using 120- or 220-grit paper). Aim for a thick peanut-butter-like consistency, as shown in photo A.

2. Once the patch is mixed, use a plastic knife to smear it into the cavity or crack, as shown in photo B. Because you'll scrape away the excess, it's okay to apply more than what's needed. Try to avoid creating any bubbles or air pockets.

3. Using a card scraper, scrape the patch and the surrounding surface of the wood, as shown in photo C. When the patch is flat, finish-sand the entire surface to conceal your work, and apply a finish.

B

Fill 'er up. Using a plastic knife, drip and spread the mix into the hole. Work quickly, or your patch may cure in the cup.

A

Making it match. You can make a matching patch with some sawdust and some 5-minute epoxy. Using the scraper to spoon in the sawdust, make the mix as thick as possible, but not crumbly.

C

Scrapes smooth, just like wood. When the patch has fully cured, a card scraper is used to smooth it out. Once the patch is flush, finish-sand the board and you're done.

Lay out the biscuit slots. **Using a scrap piece of the bed's rail as a template (instead of reading numbers off your combination square) is a good way to avoid measurement-related errors.**

Slotting the legs. **Starting with the inside face of the leg resting on your bench, use one ¼''-thick spacer to position the first biscuit slot and three to cut the second.**

Slotting the rails. **Place the rail on your bench, inside face down. With the biscuit joiner also resting on your bench, cut the first slot, then use two ¼''-thick spacers to cut the second.**

Clamp it tight. **The blue tape is used to make sure the carefully arranged slats weren't shuffled during glue-up. Put blocks under the long bars to prevent them from bowing as the clamps are tightened.**

4. You're now set to assemble the headboard and footboard. Assembling this many pieces at one time can be tricky, so try a dry run without glue. This may also be a good time to enlist an extra pair of hands to assist with positioning the slats as you clamp things together.

Glue the slats into the rails, then glue the rails to the legs, using as many clamps as necessary to pull it all together, and check for square, as shown in photo N. When clamping, it's okay if the top rail shifts down slightly, but make sure the bottom rail doesn't move from your original layout lines.

5. Give the glue at least an hour before removing the clamps. You should allow the glue at least a full day to cure completely, but at this point you can remove any drips, finish-sand the legs, and coat unfinished surfaces with oil. After two or three coats of oil, apply a coat of wax to all finished surfaces and buff to a shine.

What You'll Need

- Router with straight bit
- Four strips of plywood, ¾" by 3" by 12"
- 12" combination square
- Ten 1¼" deck screws
- Two or three 6" bar clamps

With four scraps of plywood and a handful of drywall screws, you can make a jig to cut any width of dado you need. When clamped against the board, this simple setup makes identically sized dadoes that are always square to the edge.

1. Cut the plywood strips first. Making the fences longer than the board helps guide the router's base at the beginning and end.

2. Attach the right-hand fence. Insert one screw, then use a combination square to make certain the assembly is square before driving the other two screws.

3. Clamp the partially assembled jig to your bench, adjust your router to make a ⅛"-deep cut, and make a test-cut on your front cleat. (Mark the fence and router to ensure the same side of the router is used against the same fence. Not all router bases are perfectly symmetrical; changing the orientation can affect the width of the dado.)

4. Raise the bit above the surface of the board to help position your left-hand fence. If the cut and layout lines match up, attach the back cleat.

5. Work so the bit's rotation keeps the router against your fences. Place the router against the right fence on the far side of the board and pull the router toward you. When the bit reaches the front cleat, slide the router to the left fence and push it back through the board. Remove the material in the middle of the dado with back-and-forth passes, as shown in the photo below.

Wide Dado-Routing Jig

This setup works with any router and straight bit combination.

You're ready to rout. Use the fences as guides to rout the sides of the dado, then remove the center section of the dado freehand.

For consistent-width dadoes, mark fence and router so same side of router is used against fence.

With router against right fence, rout cleat before positioning left fence.

12" to 18"

Attach right fence first.

Fences should be square to front cleat.

Adjust width to fit stock.

Back cleat

Simple assembly. **The hardware on the side rails locks into place on the legs as it's dropped into place. At this point, be careful not to twist the frame (or you could damage the fastener).**

Making the Bed

You're only a few screws and a couple of cuts away from finishing this project. Now's the time to be extra careful. An old joke claims that mistakes happen "when you start to smell the finish," but there's nothing funny about stripping or snapping that last screw or dinging up a footboard because you were too impatient to ask someone to hold open a door.

1. Mount the bed-rail fasteners to the headboard and footboard legs, as shown in the drawing at right. (If you're not using the same hardware, position the piece so the side rails attach to the legs at the same height as the bottom rail of the headboard and footboard.) Use a centering bit to drill the pilot holes, then attach the metal clips using 1" #8 wood screws.

2. Attach the side rails to the headboard. If everything was installed correctly, the two should clip together, as shown in photo A. Be careful not to twist the rails around too much at this point. The fasteners are strong enough to support the weight of the mattress and sleeper, but they can be damaged by levering the rail from side to side.

Bed-Rail Detail

Metal fasteners simplify assembly, but they should be installed precisely so that the bottom rails line up. Take a minute to see how the fasteners hook together before you start screwing them to the legs and side rails.

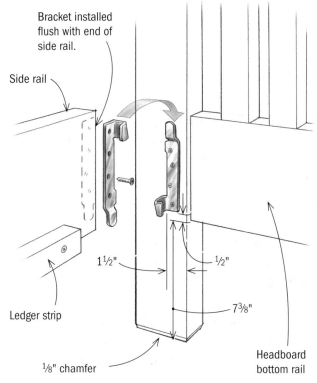

Bracket installed flush with end of side rail.

Side rail

Ledger strip

$1\frac{1}{2}$"

$\frac{1}{2}$"

$7\frac{3}{8}$"

$\frac{1}{8}$" chamfer

Headboard bottom rail

3. Cut the support planks about $\frac{1}{8}$" shorter than the distance between the side rails. The materials list specifies five or six planks, but if you're going without a box spring, you may need a few more. Alternately, if you install enough planks so there is less than a 1" gap between boards, you can use a softer futon-like mattress.

4. Complete the bed with a box spring, mattress, mattress pad, linens, and comfy pillows, and it's ready for a well-deserved rest. You probably are too.

Resources

American Clamping Corporation
50 Franklin St.
P.O. Box 399
Batavia, NY 14621
Distributors of German-made Bessey clamps.

Bench Dog Tools
9775-85th Ave. N
Maple Grove, MN 55369
www.benchdog.com
A full range of tools and accessories for routers.

Bosch Power Tools
1800 W. Central Rd.
Mt. Prospect, IL 60056
www.boschtools.com
Handheld and benchtop power tools.

Franklin International
2020 Bruck St.
Columbus, OH 43207
www.franklininternational.com
Maker of many different adhesives to match all sorts of woodworking and construction projects.

Freud America, Inc.
218 Feld Ave.
High Point, NC 27263
www.freudtools.com
Maker of blades and bits.

Garrett Wade
5389 E. Provident Dr.
Cincinnati, OH 45246
www.garrettwade.com
Retailer of fine woodworking tools.

Highland Hardware
1045 North Highland Ave. NE
Atlanta, GA 30306
www.highlandwoodworking.com
A wide range of power tools and full range of supplies.

Jamestown Distributors
17 Peckham Dr.
Bristol, RI 02809
www.jamestowndistributors.com
Specialize in fasteners of all kinds but also offers tools, sandpaper, adhesives, etc.

Kreg Jig
201 Campus Dr.
Huxley, IA 50124
www.kregtool.com
The pros' choice for pocket holes.

Lee Valley Tools Ltd.
P.O. Box 1780
Ogdensburg, NY 13669
www.leevalley.com
This company's excellent catalog is an education in woodworking.

Lie-Nielsen Toolworks, Inc.
P.O. Box 9
Warren, ME 04864
www.lie-nielsen.com
Maker of superb planes, chisels, and saws.

McFeely's
P.O. Box 44976
Madison, WI 53744
800-443-7937
www.mcfeelys.com
A wide selection of specialized and general-purpose square-drive screws for a variety of uses.

Osborne Wood Products Inc.
4618 GA Hwy. 123 N.
Toccoa, GA 30577
www.osbornewood.com
Maker of turned wooden table legs.

Porter-Cable/Delta
4825 Hwy. 45 N.
P.O. Box 2468
Jackson, TN 38302
www.portercable.com
www.deltamachinery.com
Maker of almost every type of stationary, benchtop, and handheld power tool.

Roberts Plywood
45 N. Industry Ct.
Deer Park, NY 11729
www.getwood.com
An astonishing variety of specialty plywoods, available in slightly less than half-sheets via UPS.

Rockler Woodworking and Hardware
4365 Willow Dr.
Medina, MN 55340
www.rockler.com
Mail-order source for tools, hardware, and finishes.

System Three Resins, Inc.
3500 W. Valley Highway N. Ste. 105
Auburn, WA 98001
www.systemthree.com
Maker of high-performance epoxies, paints, and varnishes, including the self-mixing gun used in the Outdoor Easy Chair.

Tried and True Finishes
14 Prospect St.
Trumansburg, NY 14886
www.triedandtruewoodfinish.com
Maker of oil/wax finishes.

Twin Oaks Lumber LLC
2345 Sinsinawa Rd.
Hazel Green, WI 53818
A fine source of domestic and exotic hardwoods and softwoods.

Waterlox Coatings Corporation
9808 Meech Ave.
Cleveland, OH 44105
www.waterlox.com
Maker of the tung oil varnish used on the Outdoor Easy Chair project.

Woodcraft
1177 Rosemar Rd.
P.O. Box 1686
Parkersburg, WV 26102
www.woodcraft.com
Catalog retailer of tools, materials, wood, and hardware.

The Woodworkers' Club
215 Westport Ave.
Norwalk, CT 06851
www.woodworkersclubnorwalk.com
Offers kits of tools and materials for projects in this book.

Woodworker's Supply, Inc.
5604 Alameda Pl. NE
Albuquerque, NM 87113
www.woodworker.com
Power and hand tools, as well as supplies.

Reading List

The Basics of Craftsmanship: Key Advice on Every Aspect of Woodworking Joinery: Shaping & Milling.
The Taunton Press, 1999. Reprints from *Fine Woodworking* magazine.

Care and Repair of Shop Machines.
John White. The Taunton Press, 2002. A complete reference for assembling, tuning, maintaining, and repairing major shop tools.

The Complete Guide to Sharpening.
Leonard Lee. The Taunton Press, 1990. An in-depth guide to sharpening all kinds of edge tools.

Encyclopedia of Furniture Making.
Ernest Joyce. Sterling Publications, 1987. This is the old standby reference for furniture makers in the United States and United Kingdom.

Fine Woodworking *Magazine.*
63 S. Main St., P. O. Box 5506, Newtown, CT 06470 www.finewoodworking.com. How to do all kinds of woodworking, from the humble to the sublime.

Great Wood Finishes: A Step-By-Step Guide to Consistent and Beautiful Results.
Jeff Jewitt. The Taunton Press, 2000. Details on finishing. This book is a necessity if you want to do anything other than oil your projects.

Identifying Wood.
R. Bruce Hoadley. The Taunton Press, 1990. Information on how to identify over 180 domestic and tropical hardwoods and softwoods.

Mastering Woodworking Machinery.
Mark Duginske. The Taunton Press, 1992. The classic guide on how to tune up and use the tablesaw and other woodworking machines.

Methods of Work: The Best Tips from 25 Years of Fine Woodworking *(Methods of Work Series).*
The Taunton Press, 2000. A collection of neat tips and tricks from 25 years of *Fine Woodworking* magazine.

The Router Book: A Complete Guide to the Machine and Its Accessories.
Pat Warner. The Taunton Press, 2001. A great guide to learning to use this essential tool.

Router Joinery.
Gary Rogowski. The Taunton Press, 1997. Jigs and setups for routing joints.

Router Magic: Jigs, Fixtures and Tricks to Unleash Your Router's Full Potential.
William H. Hylton. Reader's Digest Adult, 1999. How to do almost anything with a router.

Setting Up Shop: The Practical Guide to Designing and Building Your Dream Shop.
Sandor Nagyszalanczy. The Taunton Press, 2000. What you need to know to set up a shop right. Interesting peeks into a variety of shops.

ShopNotes *Magazine.*
www.shopnotes.com. Projects and info to get your shop in shape.

The Table Saw Book.
Kelly Mehler. The Taunton Press, revised edition 2002. A no-frills, easy to understand book on all aspects of the tablesaw.

Table Saw Magic.
Jim Tolpin. Popular Woodworking Books, 1999. Choosing, using, and mastering the tablesaw.

Tage Frid Teaches Woodworking 1 & 2: A Step-By-Step Guidebook to Essential Woodworking Technique.
Tage Frid. The Taunton Press, 1994. Classical European woodworking with an emphasis on hand tools.

Taunton's Complete Illustrated Guide to Finishing.
Jeff Jewitt. The Taunton Press, 2004. Covers all the modern and traditional techniques for coloring and finishing wood.

Taunton's Complete Illustrated Guide to Furniture & Cabinet Construction.
Andy Rae. The Taunton Press, 2001. Covers every practical technique for building furniture and cabinets, from designing the projects to installing the hardware.

Taunton's Complete Illustrated Guide to Joinery.
Gary Rogowski. The Taunton Press, 2002. Information on every practical joint used in woodworking.

Taunton's Complete Illustrated Guide to Sharpening.
Thomas Lie-Nielsen. The Taunton Press, 2004. Covers all types of sharpening equipment and methods of sharpening hand tools.

Traditional Woodworking Handtools: A Manual for the Woodworker.
Graham Blackburn. The Lyons Press, 1999. A lot of history but solid practical information for those who want to learn more about hand tools.

Understanding Wood: A Craftsman's Guide to Wood Technology.
R. Bruce Hoadley. The Taunton Press, revised edition, 2000. The standard reference on why wood does what it does.

Woodworking with the Router: Professional Router Techniques and Jigs Any Woodworker Can Use.
Bill Hylton and Fred Matlack. Reader's Digest Adult, 1999. Everything you want to know about the router. A great reference.

The Workbench Book.
Scott Landis. The Taunton Press, 1998. A cultural history of the workbench that's actually useful. Photos and plans for a variety of proven designs. Pick the one that suits your style.

Index